Introduction to phosphorus chemistry

Cambridge Texts in Chemistry and Biochemistry

GENERAL EDITORS

D. T. Elmore
Professor of Biochemistry
The Queen's University of Belfast

J. Lewis
Professor of Inorganic Chemistry
University of Cambridge

K. Schofield
Professor of Organic Chemistry
University of Exeter

J. M. Thomas
Professor of Physical Chemistry
University of Cambridge

Introduction to phosphorus chemistry

HAROLD GOLDWHITE

Professor of Chemistry, California State University, Los Angeles

CAMBRIDGE UNIVERSITY PRESS
Cambridge
London New York New Rochelle
Melbourne Sydney

CAMBRIDGE UNIVERSITY PRESS
Cambridge, New York, Melbourne, Madrid, Cape Town,
Singapore, São Paulo, Delhi, Tokyo, Mexico City

Cambridge University Press
The Edinburgh Building, Cambridge CB2 8RU, UK

Published in the United States of America by Cambridge University Press, New York

www.cambridge.org
Information on this title: www.cambridge.org/9780521297578

First published 1981
Re-issued 2011

A catalogue record for this publication is available from the British Library

Library of Congress Cataloguing in Publication data
Goldwhite, Harold.
Introduction to phosphorus chemistry.
(Cambridge texts in chemistry and biochemistry)
Includes index.
1. Phosphorus. 1. Title. 11. Series
QD181.P1G67   546'.712   79-27141

ISBN 978-0-521-22978-4 Hardback
ISBN 978-0-521-29757-8 Paperback

Contents

Contents

Contents

Preface

This book presents an introduction to the chemistry of a single element, phosphorus, as an example of a range of topics of interest in the whole of chemistry. The chemistry of phosphorus involves many of the major themes of modern chemistry – for example, bonding; physical techniques; and structure, both static and dynamic. In surveying the chemistry of phosphorus, it is necessary to cross and recross the boundaries of the various subdisciplines of chemistry that are today beginning to erode. There is an "inorganic" chemistry of phosphorus; organophosphorus chemistry is a large and expanding domain; phosphorus is an economically important element, and a vitally important element in the biosphere.

Many books have already been devoted, in whole or in part, to this important and interesting element. There are, however, few recent brief and introductory surveys, of the type this book attempts, of the major points of interest in the chemistry of this element. This book is intended for the chemist at an advanced undergraduate or graduate level who is interested in learning the fundamentals of the chemistry of phosphorus. Because an attempt is made at conciseness, the book cannot claim to be comprehensive or entirely current. The most significant developments in phosphorus chemistry through the end of June 1979 have been included. The critical bibliography and suggestions for further reading list the most useful secondary and tertiary sources in phosphorus chemistry. The primary literature is not included, because that would greatly expand the bibliography with little added utility for the intended reader of this work. Fortunately for those working in phosphorus chemistry, the major developments are rapidly made accessible in reviews and specialist reports, as indicated.

Chemists in the United States are, in the main, reluctant to adopt *Système Internationale* (SI) units. However the SI system's consistency and ease of use will certainly lead to a steady increase in its adoption by chemists. In this book I have given both SI and the more familiar cgs units where appropriate.

I carried out my original investigations in phosphorus chemistry under the direction of Dr. B. C. Saunders at Cambridge University, and later worked collaboratively in a very different area of phosphorus chemistry with Professor R. N. Haszeldine at the University of Manchester Institute of Science and Technology. I owe much to these chemists, and to my undergraduate, graduate, and postdoctoral collaborators over the years.

The present text had its origin in courses given at California State University, Los Angeles; at the University of Strasbourg; and at the National University of Mexico. I thank my colleagues and students at all three institutions for their valued contributions to the development of this material.

Finally, I thank my parents, my wife, and my children for their continuous strong support and help; their contributions to my work have been of inestimable value.

Harold Goldwhite

Los Angeles

Notes on nomenclature

The nomenclature of phosphorus chemistry is in a somewhat confused (and confusing) state. Although an "agreed" system of naming phosphorus compounds was put forward jointly by the Chemical Society of London and the American Chemical Society in 1952, it was not adopted wholeheartedly by authors or editors and in the present literature the same phosphorus compound may be given several different names. This is a fundamentally undesirable situation, but it seems unlikely to change in the near future, in view of the already existing proliferation of names. The International Union of Pure and Applied Chemistry (IUPAC) has, in its latest proposals on the nomenclature of organic compounds containing phosphorus (Tentative Nomenclature of Organic Chemistry, Section D, Bulletin No. 31, August 1973) explicitly endorsed the naming of such compounds in three different ways: (1) as substitution products of parent hydrides, for example, of PH_3, phosphine; (2) as derivatives of parent compounds, for example, H_2POH, phosphinous acid; (3) by coordination nomenclature as compounds of phosphorus, with an oxidation state affixed, if the author desires.

The proposals for phosphorus nomenclature (and that of analogous arsenic, antimony, and bismuth compounds) occupy 27 text pages in the IUPAC monograph and will not be reproduced here. Instead examples of alternative names for some common structural types will be presented, so that the reader will be able to follow not only the text of this book, but also the current literature of the subject.

Compound	Acceptable name(s) (and comments)
PH_3	Phosphine (a parent compound)
PHCH$_3$	Methyl(1-naphthyl)phosphine (alphabetic ordering of radicals)

Compound	Acceptable name(s) (and comments)

CH$_3$—CH—CH$_2$OH
       |
      PH$_2$
	2-Phosphino-1-propanol (using phosphino as a name for the —PH$_2$ substituent)

HP(CH$_2$CH$_2$COOH)$_2$	3,3'-Phosphinediyldipropionic acid (phosphinediyl is the recommended name for HP⟨ as a substituent)

$\overset{11}{\text{CH}_3}\overset{10}{\text{CH}_2}\overset{9}{\text{P}}$—$\overset{8}{\text{CH}_2}\overset{7}{\text{CH}_2}\overset{6}{\text{P}}$—$\overset{5}{\text{CH}_2}\overset{4}{\text{CH}_2}\overset{3}{\text{CH}_2}\overset{2}{\text{P}}\overset{1}{\text{CH}_3}$
      | | H
     CH$_3$ C$_2$H$_5$
	6-Ethyl-9-methyl-2,6,9 triphosphaundecane (using substitution nomenclature with the suffix-*a* for a chain compound with several phosphorus atoms)

CH$_3$COP(CH$_3$)$_2$	Acetyldimethylphosphine (acyl compounds are named as phosphine derivatives)
(C$_6$H$_5$)$_2$PLi	Lithium diphenylphosphide (the suffix-*ide* indicates the negative character of the phosphorus atom)
(C$_2$H$_5$)$_2$POCH$_3$	Diethylmethoxyphosphine (as a phosphine derivative) Methyl diethylphosphinite [as an ester of diethylphosphinous acid, (C$_2$H$_5$)$_2$POH] Diethylmethoxophosphorus (III) (coordination nomenclature)
CH$_3$PCl$_2$	Dichloro(methyl)phosphine [the (methyl) is in parentheses to avoid confusion with CHCl$_2$PH$_2$, (dichloromethyl) phosphine] Methylphosphinous dichloride Dichloro(methyl) phosphorus

Compound	Acceptable names(s) (and comments)
$C_6H_5P\!\!\begin{array}{l}\diagup OCH_3 \\ \diagdown N(CH_3)_2\end{array}$	Dimethylamino(methoxy) (phenyl)phosphine Phenyl N,N,P-trimethyl-phosphonamidite [as a derivative of phosphonamidous acid, $HP(OH)NH_2$] Dimethylamido(methoxo) phenylphosphorus
$(CH_3)(C_2H_5)(C_6H_5)(C_6H_5CH_2)P^+Cl^-$	Benzylethylmethylphenyl-phosphonium chloride (the suffix-*onium* indicates four-coordinate positive character for phosphorus)
$(C_6H_5)(CH_3)_2PO$	Dimethyl(phenyl)phosphine oxide Dimethyl(oxo)phenylphosphorane (phosphorane denotes the five-coordinate state of phosphorus with, in this example, the oxo substituent formally occupying two coordination sites; see below for other examples) Dimethyl(oxo)phenylphosphorus
$(C_6H_5)_3PCH_2$	Methylene(triphenyl)phosphorane Methylene(triphenyl)phosphorus (Triphenylphosphonio) methylide (when the contribution $P^+\!-C^-$ is to be stressed)
$(CH_3)_2PO(OH)$	Dimethylphosphinic acid Dimethyl(hydroxo)(oxo) phosphorane Dimethyl(hydroxo)(oxo) phosphorus

Compound	Acceptable names(s) (and comments)
$ClCH_2PO(OH)_2$	(Chloromethyl)phosphonic acid Chloromethyl(dihydroxo) (oxo)phosphorane Chloromethyl(dihydroxo) (oxo)phosphorus
$C_6H_5P(:O)(Cl)(OCH_3)$	Methyl phenylchlorophosphonate [ester of phenyl-chlorophosphonic acid, $C_6H_5P(O)(Cl)OH$] Methyl phenylphosphonochloridate [ester of acid chloride of phenylphosphonic acid, $C_6H_5P(O)(OH)_2$] Chloro(methoxo)(oxo) phenylphosphorus
$(C_6H_5)_3PBr_2$	Dibromotriphenylphosphorane Dibromotriphenylphosphorus
	Potassium tris(2,2'-biphenylylene) phosphate(V) (Stock system) Potassium tris(2,2'-biphenylylene)phosphate(I-) (ion-charge system)
H_2P—PH_2	Diphosphane (the old name, diphosphine, is strongly discouraged)

1   Introduction to phosphorus chemistry

1.1   Historical development

Because phosphorus compounds occur in nature both in the biosphere and in minerals, the knowledge of phosphorus compounds (though not of their nature or constitution) is part of the prehistory of chemistry. The old name, microcosmic salt, by which ammonium sodium hydrogenphosphate tetrahydrate, $NaNH_4 HPO_4 \cdot 4H_2 O$, was known in late alchemical times, indicates the significance of at least one phosphorus compound to early chemistry. The element itself was also a product of alchemy, having been first produced, by distillation of evaporated urine (presumably involving reduction of phosphate by pyrolytic carbon) by Brand, in Hamburg, in about 1674. Its most spectacular chemical property, that of chemiluminescence during aereal oxidation, led to the element being named phosphorus (from the Greek, meaning light-bearer).

The simple inorganic chemistry of phosphorus was developed by a large number of chemists during the eighteenth and nineteenth centuries. The systematic development of organophosphorus chemistry was initially undertaken by A. Michaelis during the latter part of the nineteenth century, and was also pursued by Arbuzov in the early years of the twentieth century.

Phosphorus chemistry developed rapidly in the mid-twentieth century for three main reasons. The discovery by Schrader and by Saunders of the insecticidal and toxic properties of certain esters containing phosphorus on the eve of World War II set in train intensive studies of this area of phosphorus chemistry, and the economic significance of phosphorus-based pesticides continues to stimulate much work. The utility of polyphosphates as builders and sequestering agents in detergents also led to rapid development of this area of phosphorus chemistry. And finally, the resurgence of inorganic chemistry in the 1950s involved, in large measure, organophosphorus compounds as ligands in many of the novel coordination compounds under study.

1

1.2   Phosphorus in the economy

In 1840 Liebig pointed out that phosphorus is an essential plant nutrient, and indicated the utility of bonemeal in agriculture. By the late 1840s processes for utilizing mineral phosphates as fertilizers had been developed in Britain. This is still by far the predominant use of phosphorus in the economy. Chart 1 gives an approximate breakdown for the United States in the early 1970s of the major classes of industrial phosphorus intermediates, and typical end uses of products containing phosphorus.

1.3   Phosphorus in the ecology

Phosphorus is, on a cosmic basis, a moderately abundant element, among the 20 most abundant in the solar system, for example. In the lithosphere the average amount of phosphorus is about 0.2%; lunar samples and meteorites average about 0.1% phosphorus. In the lithosphere there are many different minerals containing phosphorus, but the form in which the element is found

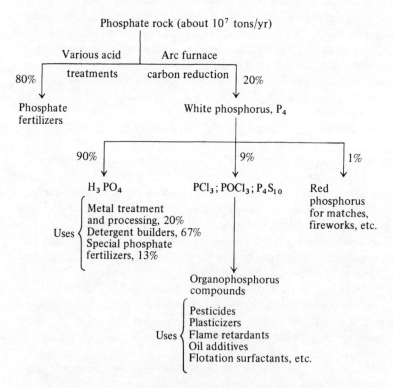

Chart 1. Phosphorus in the economy (figures for U.S. production, early 1970s).

is virtually without exception as phosphate, conventionally written as $PO_4{}^{3-}$ (formal oxidation state V; coordination number 4). The most common single mineral containing phosphorus is fluoroapatite, formula $Ca_5(PO_4)_3 F$, but this is an idealized formulation; the class of apatites has the composition $Ca_5(PO_4, CO_3)_3 (F, Cl, OH)$.

Because phosphorus is so abundant in the earth's crust, it is perhaps not surprising to find that the element has biological functions. These can be divided into two different types. In the first type, phosphorus – in the form of inorganic, primarily calcium, phosphates – is the major structural material in vertebrate bone.

In the second type, phosphate esters provide linkages in biologically important systems. Thus the major energy-storage and -transfer mechanisms in all living systems involve the synthesis and breakdown of phosphate ester linkages such as those present in adenosine diphosphate, ADP, and adenosine triphosphate, ATP (1.1), whereas the storage and transfer of coded genetic information involve nucleic acids (DNA, RNA) that are diesters of phosphoric acid [represented schematically in (1.2)]. Some of the fundamental chemistry of phosphate esters and polyphosphates will be considered later (Section 6.3).

In nature, the phosphorus cycle is different in one respect from the cycles of most other elements of biochemical significance (C, N, S, etc.). Throughout the cycle phosphorus stays in its highest oxidation state, as P(V). Many of the details of the cycle are still obscure, particularly their quantitative aspects, but its main outlines are as indicated in Chart 2.

(1.1)   Structures of un-ionized forms of adenosine diphosphate (ADP, $n = 1$ and adenosine triphosphate (ATP, $n = 2$)

It is believed that in some ecosystems phosphorus is a limiting basic nutrient. Thus when detergents containing large amounts of phosphate builders became popular in the 1950s, and large amounts of treated sewage containing phosphates began to be released into rivers and lakes, there were severe problems of eutrophication because of an upsurge in growth of algae and other primitive plants. It is only fair to add, however, that the culprit in these examples of ecological imbalance has not always been firmly identified; some

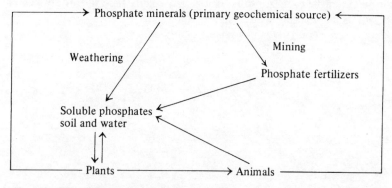

(1.2) Schematic of phosphate ester linkage in a typical nucleic acid fragment (shown un-ionized)

Chart 2. The phosphorus cycle in nature.

scientists believe that an associated increase in nitrogenous nutrients led to the rapid algal growth. Because of the possibility of phosphates being implicated, however, most detergent manufacturers have reduced the proportion of phosphate builders in detergents.

When considering the chemistry of those main group elements clustered towards the center of the periodic table (groups III, IV, and V), oxidation state is not a particularly useful method of organizing information. For instance, PF_3 (phosphorus in formal oxidation state +3), PHF_2 (phosphorus in formal oxidation state +1), and PH_3 (phosphorus in formal oxidation state -3) would be separated by an oxidation-state classification, yet the similarities in their structures and chemistry are arguably more important than their differences. Consequently, in this chapter the survey of types of phosphorus compound has been made using coordination number (C.N.) as the unifying principle.

2.1   C.N.1

C.N.1 is a very unusual coordination number for phosphorus, but the few compounds in which it is known are of considerable theoretical significance (see Section 4.3). For example, when elemental phosphorus vapor is heated, the gas-phase molecule P_2 is detectable spectroscopically (2.1). Similarly, the molecule $PN(g)$ is detectable when phosphorus nitrides are heated.

$$P(s) \rightleftharpoons P_4(g) \rightleftharpoons 2P_2(g) \rightleftharpoons 4P(g)$$
(any form: red, white, black)

(2.1)   Equilibria in elementary phosphorus

   The best-studied molecule of this kind, HCP, has been obtained both in the gas phase and in solution. It is produced when phosphine is passed through a carbon arc. In condensed phases it polymerizes rapidly above 170 K. Spectroscopic study has demonstrated the nature of the carbon–phosphorus multiple bond in HCP as that of a triple bond involving $P(3p)-C(2p)\pi$ overlap: $H-C \vdots P \colon$.

2.2   C.N.2

C.N.2 is also an unusual coordination state for phosphorus, although many more compounds with this coordination are known than with C.N.1. For example, reactions between certain cyclopolyphosphines (in which phosphorus has C.N.3) and either tertiary amines or tertiary phosphines give rather labile compounds with one phosphorus atom of C.N.2:

(2.2) $(CF_3P)_{4,5} \xrightarrow[\text{(E = N or P)}]{R_3E} CF_3PER_3$

There is also an extensive chemistry of heterocyclic analogs of benzene containing phosphorus atoms, in which phosphorus has C.N.2; for example, the parent phosphabenzene has recently been prepared:

(2.3)

Other two-coordinate phosphorus systems include the phosphacyanins (historically the first of this coordination number to be prepared),

(2.4)

some cationic compounds such as $[(CH_3)_2N]_2P^+AlCl_4^-$; the very stable silylated phosphazenes,

(2.5) $[(CH_3)_3Si]_2NPF_2 + [(CH_3)_3Si]_2NLi \longrightarrow [(CH_3)_3Si]_2NP{:}NSi(CH_3)_3$

and some novel stable compounds containing phosphorus–carbon double bonds:

(2.6)

The stability of the 2-C.N. compounds shown in (2.5) and (2.6) is in part due to shielding of the phosphorus atoms by the bulky groups present.

So far there are few structural studies on these 2-C.N. compounds. The CPC skeleton of the phosphacyanins (2.4) is linear, whereas in a phosphazene of the type shown in (2.5) the NPN angle is about 105°.

2.3   C.N.3

C.N.3 is one of the most common coordination numbers for phosphorus. An enormous range of compounds of this type, PXYZ, is known, where X, Y, and Z may be H, halogen, organic substituents, OR, NR_2, PR_2, SR, or SiR_3 groups (there are limitations on allowable combinations of groups because of mutual reactivity). The common forms of elemental phosphorus – namely, white phosphorus, P_4, and the polymeric amorphous red phosphorus (2.7) – contain three-coordinate phosphorus atoms.

(2.7)

P_4
a tetrahedral molecule

red P
polymeric and amorphous

Chlorination of elemental phosphorus produces phosphorus trichloride, PCl_3, another C.N.3 compound, which is the starting material used in preparation of most other C.N.3 products.

The geometries and structures of many C.N.3 phosphorus compounds have been determined. The almost invariant geometry is that of a trigonal pyramid (2.8), and the angles at phosphorus are normally less than 100°. The stereo-

(2.8)

chemical and bonding implications of this structure will be enlarged upon later (section 5.2).

The first exception to the general rule of trigonal pyramidal geometry for C.N.3 was recently reported. The silylated phosphorus–nitrogen compound shown as (2.9), which is an isolable analog of a monomeric metaphosphate (cf. Section 6.2.3), is reported to have a planar trigonal arrangement of the three nitrogen atoms around phosphorus.

$$[(CH_3)_3Si]_2NP{:}NSi(CH_3)_3 + (CH_3)_3SiN_3$$

$$\longrightarrow N_2 + [(CH_3)_3Si]_2NP{\Big\langle}\genfrac{}{}{0pt}{}{NSi(CH_3)_3}{NSi(CH_3)_3}$$

(2.9)   Trigonal planar coordination about P

2.4   C.N.4

C.N.4 is the most common coordination type for phosphorus compounds, the geometry being that of a tetrahedral arrangement of groups around a central phosphorus atom. An extremely wide range of groups can be attached to phosphorus in this coordination type. Terrestrial naturally occurring phosphorus compounds, both in the lithosphere and the biosphere, are of this coordination number and geometry; (2.10) shows in general terms some of the variations possible on the theme of four coordination. Specific compounds will be discussed in detail in Chapter 6.

R_4P^+ $R = H, CH_3, C_6H_5, OR', NR_2', F, Cl$, etc.
Phosphonium
cation

$R_3P{:}E$ $E = O, S, Se, NR', CR_2'$, or a Lewis acid,
        e.g., BH_3, BF_3, or a wide range of metallic compounds
        $R = H, CH_3, C_6H_5, OR', NR_2', SR, F, Cl$, etc.

(2.10)   Some phosphorus compounds of C.N.4

2.5   C.N.5

C.N.5 is not uncommon for phosphorus, with a wider and wider range of examples being reported. The simplest compounds of this type are the gaseous pentahalides PF_5 and PCl_5. (The gaseous pentahydride, PH_5, phosphorane, is as yet unknown). There are now many organophosphorus compounds known with C.N.5. Two idealized geometries have been used to describe compounds in this group. The more common one is trigonal bipyramidal (t.b.p.), such as is found in gaseous phosphorus pentafluoride (2.11); the less common one is square pyramidal (s.p.), and an example of this geometry is also shown in (2.11). An extensive discussion of the stereochemistry of C.N.5 is given in Chapter 7.

It should be mentioned that in the chemistry of phosphorus (as in many other areas of chemistry), stoichiometry is a very shaky guide to constitution or structure. Phosphorus pentachloride is an instructive example. In the gas phase, electron diffraction shows PCl_5 to be t.b.p., a conclusion that is supported by vibrational spectroscopy of the gas phase. Vibrational spectral

t.b.p.   s.p.

(2.11)   Geometry of C.N.5

studies of solutions of PCl_5 in benzene or carbon disulfide indicate that it retains its t.b.p. structure (D_{3h} symmetry) in these solvents, but similar studies of dilute solutions in acetonitrile (a much more polar solvent) indicate that PCl_5 is present as the ionized species $PCl_4^+PCl_6^-$. Spectroscopic and cryoscopic studies of solutions in carbon tetrachloride indicate the presence of a dimer, P_2Cl_{10}, of unknown structure [although the plausible chlorine bridged structure (2.12) can be suggested]. Finally, X-ray diffraction studies of crystalline PCl_5 show the presence in the structure of equal numbers of tetrahedral PCl_4^+ ions and octahedral PCl_6^- ions.

(2.12)   A possible structure for P_2Cl_{10}

2.6   C.N.6

C.N.6 is again a less common C.N. for phosphorus, although there are indications that there is potentially a wide range of compounds available with this coordination. The geometry for all examples studied to date is octahedral. The two best-known compounds of C.N.6 are the anions PF_6^- and PCl_6^-. Organoderivatives have also been prepared, for example, in the elegant synthesis of a chiral derivative shown in (2.13).

2.7   Reactions by bond types

To complete this general survey of phosphorus chemistry, and before going on to a detailed discussion of the physical properties and chemistry of the

$$PCl_4{}^+ PCl_6{}^- \xrightarrow[\text{excess}]{\overset{\text{Li}}{\underset{\text{Li}}{\bigcirc}}} \left(\bigcirc\right)_2 P^+ \quad \left(\bigcirc\right)_3 P^-$$

tetrahedral   octahedral and
chiral; has
been resolved

(2.13)

various classes of phosphorus compounds, it will be helpful to provide a brief general discussion of the chemistry of some commonly occurring phosphorus systems. Because the later discussion, in Chapters 5, 6, 7, and 9, is organized in terms of particular coordination numbers or specific reagent types, a different approach is adopted in this section and reactions are grouped according to the type of bond to phosphorus.

2.7.1 Reactions of P-H bonds

Phosphorus and hydrogen have very similar electronegativities, and consequently polar reactions are not much favored at the P-H bond. Hydrogen on phosphorus is only weakly acidic in C.N.3 compounds, although it can be abstracted by alkali metals or metal alkyls:

(2.14)
$$PH_3 + Na \xrightarrow{NH_3(l)} NaPH_2 + \tfrac{1}{2} H_2$$

$$(C_6H_5)_2PH + n\text{-BuLi} \longrightarrow (C_6H_5)_2PLi + n\text{-BuH}$$

In C.N.4 compounds, hydrogen on phosphorus is substantially more acidic and can be abstracted by alkoxide ions, as well as the more basic reagents mentioned above:

(2.15) $(CH_3O)_2P(:O)H + CH_3O^- \rightleftharpoons (CH_3O)_2P(:O)^- + CH_3OH$
dimethyl phosphite

If this last reaction is carried out in CH_3OD, deuterium is rapidly incorporated into the dimethyl phosphite.

Dialkyl phosphites readily form salts by reaction with alkali metals, and the alkylation of these salts is an important preparative method in phosphorus chemistry (Michaelis-Becker reaction):

$$(C_2H_5O)_2P(:O)H \xrightarrow{\text{Na}} (C_2H_5O)_2P(:O)Na$$

(2.16)

$$\xrightarrow{n C_3H_7Br} (C_2H_5O)_2P(:O)n\text{-}C_3H_7$$

The P–H bond of phosphites and phosphines can be added, with base catalysts, to unsaturated compounds:

$$(C_2H_5O)_2P(:O)H + CH_2:CHCN \xrightarrow{\text{base}} (C_2H_5O)_2P(:O)CH_2CH_2CN$$

(2.17) $(C_6H_5)_2PH + CH_2:CHP(:S)(CH_3)_2$

$$\xrightarrow{t\text{-BuOK}} (C_6H_5)_2PCH_2CH_2P(:S)(CH_3)_2$$

P–H bonds can also be attacked under homolytic conditions, and radical substitution and addition reactions are well characterized:

$$PH_3 + \text{[cyclohexene]} \xrightarrow[\substack{\text{radical} \\ \text{initiators}}]{h\nu} cC_6H_{11}PH_2 + (cC_6H_{11})_2PH + (cC_6H_{11})_3P$$

(2.18)

$$CHF_2CF_2PH_2 + 2Cl_2 \xrightarrow[\text{temp.}]{\text{low}} CHF_2CF_2PCl_2 + 2HCl$$

P–H bonds in C.N.3 compounds are readily oxidized by molecular oxygen, but these reactions are usually vigorous and uncontrollable.

2.7.2 Reactions of P–C bonds

The phosphorus–carbon bond is strong and not easily cleaved. Generally, P–C bonds are preserved even under vigorous reaction conditions. However, if the carbon group on phosphorus is electron attracting and can form a relatively stable carbanion, there are some reactions that can lead to P–C bond cleavage, for example,

(2.19) $(C_6H_5)_3P + 2Li \longrightarrow (C_6H_5)_2PLi + C_6H_5Li$

Bases cleave P–C bonds in phosphonium salts, the group cleaved being the one that best accommodates the developing negative charge (see Section 6.3.1):

(2.20) $C_6H_5CH_2P(CH_3)_3^+ \xrightarrow[H_2O]{HO^-} C_6H_5CH_3 + (CH_3)_3PO$

2.7.3 Reactions of P–N bonds

The P–N bond is moderately polar, and is susceptible to attack by nucleophiles. The best-documented reactions are base hydrolyses of phosphoramides and alcoholyses of phosphorous amides:

$$C_6H_5P(:O)[N(CH_3)_2]_2 \xrightarrow[H_2O]{HO^-} C_6H_5P(:O)(OH)_2 + 2(CH_3)_2NH$$

(2.21)

$$[(CH_3)_2N]_3P + C_2H_5OH \longrightarrow [(CH_3)_2N]_2POC_2H_5 + (CH_3)_2NH$$

Phosphorous amides are also easily cleaved by anhydrous hydrogen halides, yielding P-halogen compounds:

(2.22) $\quad (C_2H_5)_2NPF_2 + 2HBr \longrightarrow PBrF_2 + (C_2H_5)_2NH_2^+Br^-$

2.7.4 Reactions of P-P bonds

The P-P bond is not very strong and is susceptible to oxidation and to reduction:

$$(CH_3)_4P_2 \xrightarrow[H_2O]{KMnO_4} 2(CH_3)_2P(:O)OH$$

(2.23)

$$\xrightarrow[\text{ether}]{LiAlH_4} 2(CH_3)_2PH$$

With thermal initiation, some tetraalkyl diphosphines add to alkenes, apparently in a homolytic manner:

(2.24) $\quad (CF_3)_2PP(CF_3)_2 + CH_2:CH_2 \longrightarrow (CF_3)_2PCH_2CH_2(CF_3)_2$

2.7.5 Reactions of P-O bonds

The P-O bond is a polar bond and in phosphoric and phosphorous esters is readily attacked by bases:

$$C_6H_5P(:O)(OCH_3)_2 \xrightarrow[H_2O]{OH^-} C_6H_5P(:O)(OH)_2 + 2CH_3OH$$

(2.25) $\quad C_2H_5P(OC_2H_5)_2 \xrightarrow{H_2O} C_2H_5PO_2H_2 + 2C_2H_5OH$
$$\text{N.B.: 4 C.N. } C_2H_5P(:O)(H)(OH)$$

In phosphoric esters, P-O alkyl groups are convertible to P-Cl groups by reaction with PCl_5:

(2.26) $\quad CH_3P(:O)(OCH_3)_2 + 2PCl_5 \longrightarrow CH_3P(:O)Cl_2 + 2CH_3Cl + 2POCl_3$

2.7.6 Reactions of P:O bonds

The P:O bond is very stable, and its only noteworthy reaction (apart from its potential function as a Lewis base via lone-pair electrons on oxygen) is reduction by $LiAlH_4$, or by a variety of silicon compounds:

(2.27)
$$CH_3P(:O)(OCH_3)_2 \xrightarrow{\text{LiAlH}_4} CH_3PH_2$$
$$C_6H_5P(:O)(C_2H_5)_2 + Si_2Cl_6 \longrightarrow C_6H_5P(C_2H_5)_2 + Cl_3SiOSiCl_3$$

2.7.7 Reactions of P:S bonds

The P:S bond is readily desulfurized with reducing agents to yield C.N.3 phosphorus compounds (2.28). This is often a useful preparative process.

(2.28)
$$ClCH_2P(:S)Cl_2 + PhPCl_2 \longrightarrow ClCH_2PCl_2 + PhP(:S)Cl_2$$
$$C_6H_5P[CH_2CH_2P(:S)(CH_3)_2]_2 \xrightarrow{\text{LiAlH}_4} C_6H_5P[CH_2CH_2P(CH_3)_2]_2$$

2.7.8 Reactions of P-halogen bonds

Phosphorus-halogen bonds are polar and are readily attacked by nucleophilic reagents and by metals. Reactions of P-halogen bonds are among the most

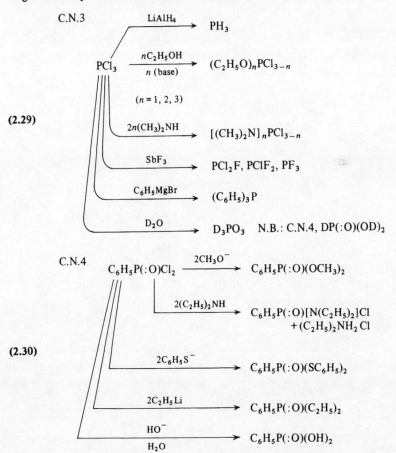

versatile and useful preparative reactions in phosphorus chemistry, and some examples for both C.N.3 and C.N.4 compounds are given in (2.29) and (2.30).

Phosphorus–halogen bonds will also undergo free-radical addition reactions to alkenes, for example,

(2.31) $PBr_3 + H_2C{:}CH_2 \xrightarrow{\text{heat}} BrCH_2CH_2PBr_2$

3 Physical methods in phosphorus chemistry

Because this book is a brief guide to phosphorus chemistry, it would be inappropriate to discuss the general principles and elementary applications of the various physical methods covered. It is assumed that the reader is already familiar with these, or can obtain the necessary background from references given in the bibliography (Appendix I). This chapter presents only those aspects of particular physical methods that are of special utility in the area of phosphorus chemistry. For collections of data, experimental methods, or individual spectra, the bibliography (Appendix I) provides a detailed description of the scope and level of each work, and an assessment of its utility.

3.1 Vibrational spectroscopy

Infrared and Raman spectroscopy have been used extensively in the exploration of structure and bonding in phosphorus compounds. In simple molecules (containing, say, not more than 8 or 10 atoms) a full vibrational assignment can often be made with confidence, and this may lead to a choice among alternative geometries or constitutions (cf. the discussion on phosphorus pentachloride in Section 2.5). In complex molecules, full assignments are made less often, but the use of group frequency arguments can often aid in structure determination. Extensive and critical compilations are available of group frequencies in phosphorus compounds. In this section some examples of the use of group frequencies in assigning structures to phosphorus compounds will be presented, and some of the more useful spectra/structure correlations will be examined.

In the reaction between ethanol and phosphorus trichloride, in the absence of base, a product of the molecular formula $C_4H_{11}PO_3$ is isolated; the stoichiometry of the full reaction is given in (3.1):

(3.1) $3C_2H_5OH + PCl_3 \longrightarrow C_4H_{11}PO_3 + C_2H_5Cl + 2HCl$

Base-catalyzed hydrolysis of the compound $C_4H_{11}PO_3$ yields 2 moles of eth-

15

anol and 1 mole of phosphorous acid, H_3PO_3. Clearly the compound is the diethyl ester of phosphorous acid, $(C_2H_5O)_2POH$. A constitutional question still remains, however, which can be solved by infrared observations. Is the phosphorus atom in this molecule of C.N.3˙or of C.N.4 (see 3.2)?

(3.2)

$$\begin{array}{ccc} C_2H_5O \\ \diagdown \\ P-O & \quad\quad & C_2H_5O \diagdown \;\; O \\ \diagup \quad\quad \diagdown & & P \\ C_2H_5O \quad\quad H & & C_2H_5O \diagup \;\; \diagdown H \end{array}$$

C.N.3   C.N.4

The infrared spectrum of the product shows strong bands at 2425 cm⁻¹, within the group frequency range normally assigned to ν_{PH} (2460-2240 cm⁻¹) and at 1255 cm⁻¹, within the range normally assigned to $\nu_{P:O}$ (1415-1087 cm⁻¹). It shows no band in the region 2725-2525 cm⁻¹, where we would expect, by group frequency arguments, to find absorption due to ν_{OH} of a P—O—H group. Consequently, the product must be assigned the C.N.4 structure. [Incidentally, similar conclusions may be drawn about most compounds of general formula R_1R_2POH, where R = alkyl, aryl, alkoxy, dialkylamino, etc. Virtually all of these occur in the C.N.4 structure. A notable exception is the compound $(CF_3)_2POH$, which has been shown, by infrared and magnetic resonance observations, to have phosphorus of C.N.3.]

One of the diagnostically most useful absorptions is that of the phosphoryl group, —P=O, in C.N.4 compounds. The absorption is usually one of the strongest bands in the infrared spectrum, and occurs within the range 1415-1080 cm⁻¹, although most compounds absorb in the narrower range of 1300-1200 cm⁻¹. Strong hydrogen bonding or association, for example, in compounds of the type $R_1R_2P(:O)(OH)$, or coordination of the phosphoryl group to a Lewis acid such as I_2 or a metal cation, results in substantial shifts of the absorption to lower frequency.

The systematics of the position of the absorption band have been widely studied, and for a large number of compounds an empirical equation (3.3) has been derived that fits the position of the band quite accurately. For

Π constants					
R	Π	R	Π	R	Π
CH_3	2.1	C_6H_5	2.4	OCH_3	2.9
NR_2	2.2	H	2.5	Br	3.1
SR	2.4	Cl	2.75	F	3.1

(3.3)   Empirical fitting of $\nu_{P:O}$

$R_1R_2R_3P{=}O$,

$$\nu_{P=O}\ (cm^{-1}) = 930\ cm^{-1} + 40\ cm^{-1}\ (\Sigma\Pi)$$

where $\Sigma\Pi$ is the sum of the empirical values for R_1, R_2, and R_3.

In some phosphoryl compounds, however, a splitting of $\nu_{P:O}$ has been observed. Thus, for $Cl_3P{:}O$, a single sharp band for $\nu_{P:O}$ is seen at 1303 cm^{-1}; in the related $(CH_3O)(Cl)(Cl)P{:}O$, two bands of equal intensity are seen at 1322 cm^{-1} and 1300 cm^{-1}. Such splittings have been explained by postulating the presence of an equilibrium mixture of rotamers (3.4), each with its distinct $\nu_{P:O}$.

(3.4) Rotamers of $CH_3OP({:}O)Cl_2$

3.2 Magnetic resonance

Phosphorus is a monoisotopic element; the species $^{31}_{15}P$ is 100% abundant in naturally occurring phosphorus and has spin quantum number $I = \frac{1}{2}$. Consequently, phosphorus is detectable in nuclear magnetic resonance (nmr) spectroscopy both by direct study of the phosphorus nucleus and by its coupling to other magnetic nuclei such as ^1H, ^{13}C, or ^{19}F. In electron spin resonance (esr) spectroscopy, phosphorus, with $I = \frac{1}{2}$, gives hyperfine coupling, which can be very helpful in spectral interpretation.

3.2.1 ^{31}P nmr

Because of a relatively low magnetic moment and magnetogyric ratio, the relative sensitivity for phosphorus nuclei in a continuous-wave (conventional) spectrometer is only 0.066 that for protons at the same field strength. This was a major problem for phosphorus nmr, but the advent of pulsed Fourier transform spectroscopy has improved sensitivity greatly, thus allowing study of compounds of limited solubility or difficult availability.

Chemical shifts in ^{31}P nmr are normally referenced to 85% H_3PO_4 (*aq*.) as external standard; $P_4O_6(l)$ and $(CH_3O)_3P$ are sometimes used as secondary external standards. Because the range of chemical shifts for ^{31}P is very large (> 700 ppm), the diamagnetic correction is rarely made. In (3.5) some typical ^{31}P chemical shift regions are indicated for both C.N.3 and C.N.4 compounds.

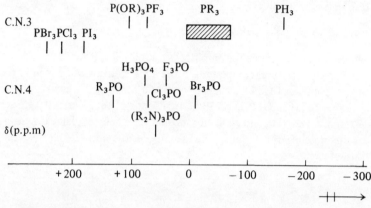

(3.5) ^{31}P chemical shifts

Attempts to interpret trends in ^{31}P chemical shifts using additivity relationships or simple hybridization arguments have not been very successful. Even
elaborate quantum mechanical calculations, which have successfully rationalized ^{31}P chemical shifts in molecules of the types PZ_3 and Z_3PO (where all
Z's are the same in a given molecule), have been of little success in systems
with mixed ligands around phosphorus. The best advice that can be given to
the beginner in this area is to use analogy as creatively as possible; tabulations
of several thousand ^{31}P chemical shifts in a wide range of compounds are
readily available.

Because they involve only phosphorus nuclei, it is convenient to discuss
phosphorus–phosphorus coupling constants in this section (although, in practice, many J_{PP} values are obtained by observation of other nuclei). The value
of $^1J_{PP}$ is extremely variable in magnitude. Some values are given in (3.6).
One-bond couplings are believed to be dominated by the Fermi contact term,
which depends on the s electron density at the nuclei. A hybridization argument can rationalize some of the trends apparent in (3.6). Thus $^1J_{PP}$ is greater
in compounds where C.N.4 nuclei are directly bonded than in those where
C.N.3 nuclei are directly bonded. In the C.N.3 compounds (cf. Section 4.1),
there is less s character in the bonds at phosphorus, because in these compounds characteristic angles at phosphorus are less than 100°, indicating much
p character in the bonds. In the C.N.4 compounds, where angles at phosphorus approach the tetrahedral, there must be more s character in the phosphorus–phosphorus bond, and hence $^1J_{PP}$ is greater. However, it has also been
shown that conformational factors may affect $^1J_{PP}$. In C.N.3 compounds the
magnitude of $^1J_{PP}$ has been shown to increase as the dihedral angle between
the directions assigned to the lone-pair electrons on adjacent phosphorus
atoms increases from 0 to 180°

Compound	J	Compound	J
$Bu_2{}^tPPBu_2{}^t$ (gauche)	-451	$(CH_3)_2P(S)P(S)(CH_3)_2$	$(\pm)\,18.7$
$(CH_3)_3P{:}PCF_3$	-445		$+465$
$(CH_3)_2PP(CH_3)_2$	-180		
H_2PPH_2	-108		

(3.6) Some values of ${}^1J_{PP}$ (Hz)

3.2.2 ${}^1H$ nmr

Proton nmr, because of its wide availability and sensitivity, has been much used in phosphorus chemistry. The most useful information has been derived from coupling constants. The value of ${}^1J_{PH}$, for example, is large and diagnostically useful. Some characteristic values are given in (3.7), and it can be seen that ${}^1J_{PH}$, like ${}^1J_{PP}$, increases as the amount of s character in the P–H bond increases. Unfortunately, efforts to quantify this trend have not been successful so far.

Compound	${}^1J_{PH}$ (Hz)	Compound	${}^1J_{PH}$ (Hz)
$PH_2{}^-$	138	PF_4H	1084
PH_3	188	PF_5H^-	955
$PH_4{}^+$	548		
$(RO)_2P(:O)H$	700		

(3.7) Some ${}^1J_{PH}$ values (Hz)

Longer-range couplings to phosphorus have attracted much interest because of their potential utility in conformational analysis. The geminal ${}^2J_{PCH}$ varies over a considerable range and has been applied to conformational analyses of phosphines. The variation of this coupling constant with a defined dihedral angle involving the direction assigned to the lone pair at phosphorus is shown in (3.8). In the cycloaddition reaction shown in (3.9), two diastereomeric phosphines are produced. The value of ${}^2J_{PCH}$ of $+22.7$ Hz in isomer A indicates a dihedral angle α of near zero in this isomer and so establishes the cis, cis, cis relationship of the methyl groups.

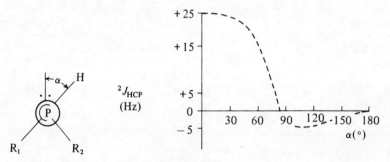

(3.8) Variation of $^2J_{PCH}$ with dihedral angle

Similar Karplus-like dependencies have been deduced for $^3J_{PCCH}$ and for $^3J_{POCH}$ (the latter being potentially useful in the conformational analysis of biochemically significant phosphorus compounds such as sugar phosphates). A cautionary note must, however, be sounded. Changes in coordination num-

(3.9)

40% A; 60% B

isomer A; $^2J_{PCH} = +22.7$ Hz

ber, charge, or substituent type at phosphorus can all affect coupling constants, and so conformational effects can be assessed only after these other effects have all been taken into consideration.

3.2.3 ^{13}C nmr

The routine availability of ^{13}C nmr spectra by Fourier transform spectroscopy has led to a rapid increase in the volume of ^{13}C chemical shift and $^{31}P-^{13}C$ coupling constant data for phosphorus compounds. There are no general principles that have yet emerged as clearly as for $^1H-^{31}P$ coupling constants, but it is clear that ^{13}C coupling to phosphorus is very sensitive to the nature of the phosphorus atom, and that spatial relationships play an important role. Some representative data are given in (3.10).

	$\delta(C_1)$	$^1J_{CP}$	$\delta(C_2)$	$^2J_{CCP}$	$\delta(C_3)$	$^3J_{CCCP}$
PMe_3	14.3	-13.6				
PEt_3	19.5	14.0	10.3	13.8		
$PPr_2{}^nPh$	31.7	12.8	20.1	15.4	16.1	12.2 (for Pr^n)
$PBu_3{}^n$	29.3	13.8	28.6	14.8	25.4	11.1

	$^1J_{CP}$		$^1J_{CP}$
$MePCl_2$	-45	$MeP(O)(OEt)_2$	$+143$
$MeP(S)Cl_2$	81		
$MeP(O)Cl_2$	104		
$MeP(O)F_2$	147		

(3.10) ^{13}C nmr data δ values in parts per million from Me_4Si; J values in hertz

3.2.4 Electron spin resonance

Electron spin resonance (esr) has been applied to the elucidation of the structures of a number of transient intermediates in reactions of phosphorus compounds. For example γ radiolysis of ammonium hexafluorophosphate, $NH_4{}^+PF_6{}^-$, leads to observation of an esr signal attributed to the $PF_4^{\cdot}$ radical that is considerably distorted from tetrahedral symmetry; the hyperfine splitting due to phosphorus in this radical is 133 mT (1330 G).

In the photolysis of di-*tert*-butyl peroxide in phosphine (PH_3) at 178 K, radical intermediates have been seen and identified. For instance, one such radical shows a doublet splitting of 62.7 mT (627 G), which must be due to phosphorus, and which is further split into a large doublet (13.96 mT) of 1:3:1 triplets (splitting 1.08 mT). This signal is plausibly assigned to a phosphoranyl radical $(CH_3)_3CO\overset{\cdot}{P}H_3$ in which one of the protons attached to phosphorus is in a unique position and interacts much more strongly (13.96 mT) with the unpaired electron than the remaining two equivalent protons (1.08 mT). A plausible structure for such a radical is C.N.5 t.b.p., as shown in (3.11). The

(3.11)   Photolysis of $(tBuO)_2$ in PH_3

line shapes for this radical are strongly temperature dependent above 193 K, and this has been interpreted in terms of a rapid ligand rearrangement within the radical; as we shall see (Section 7.2), this is a common feature of C.N.5 phosphorus compounds.

Recently some stable 2-coordinate phosphino radicals have been described in which radical stability has been enhanced by attaching very bulky groups

$$\{[(CH_3)_3Si]_2CH\}_2PCl + \quad \xrightarrow[\substack{toluene \\ 300\ K}]{h\nu} \quad \{[(CH_3)_3Si]_2CH\}_2P\cdot$$

$$a(P) = 9.63\ mT$$
$$a(H) = 0.64\ mT$$

(3.12)   Stable phosphino radicals

to phosphorus, as shown in (3.12).

3.3   Mass spectra and ion-cyclotron resonance

3.3.1   *Mass spectra*

Because phosphorus is monoisotopic, mass spectra (ms) of phosphorus compounds do not contain the useful cluster of peaks that make it easy, for example, to identify fragments containing chlorine (^{35}Cl and ^{37}Cl) in chlorinated compounds. Many mass spectral fragmentation studies of phosphorus compounds have been made, but the results are useful mainly as a fingerprinting method. Few useful generalizations have emerged. Parent ions are normally fairly intense, and rearrangements are very common, especially hydrogen migration to phosphorus. Species that may contain multiple bonds to phosphorus are also often seen. The fragmentation patterns of trimethylphosphine, $(CH_3)_3P$, and triethylphosphine, $(C_2H_5)_3P$, are given in (3.13) as examples of these general rules.

$$(CH_3)_3P^{+\cdot} \longrightarrow H + (CH_3)_2PCH_2^{+\cdot} \longrightarrow C_3H_5^+ + PH_3$$

$$CH_3 + (CH_3)_2P^{+\cdot} \longrightarrow (CH_3)_2PC^{+\cdot} + H_2$$

$$H_2 + (CH_2)_2P^+ \qquad (CH_2)_2PC^{+\cdot} + H_2$$

may be $\underset{\diagdown\quad\diagup}{CH_2\!-\!CH_2} \atop P+\cdot$ or $H_2C\!=\!P\!=\!CH_2 \atop +\cdot$

(cont.)

$$(C_2H_5)_3P^{+\cdot} \longrightarrow C_2H_4 + (C_2H_5)_2PH^{+\cdot}$$

$$\longrightarrow C_2H_4 + (C_2H_5)PH_2^{+\cdot} \longrightarrow C_2H_4 + PH_3^{+\cdot}$$

(3.13)   Mass spectral fragmentation of Me_3P and Et_3P

3.3.2   Ion-cyclotron resonance

The technique of ion-cyclotron resonance (icr) has been applied primarily to simple C.N.3 compounds to examine ion reactions in the gas phase, and to determine, via proton-transfer reactions, the gas-phase basicities of phosphines. Results from the latter aspect will be discussed in Section 5.2. Ionic reactions in the gas phase among simpler phosphines are quite complex and tend to produce polyphosphorus compounds, especially as pressures increase. Some examples are presented in (3.14).

Reaction		Rate constant, K_2 (10^{-16} m^3 molecule^{-1} sec^{-1}) at 300 K	
$PH_3^{+\cdot} + PH_3$	$\longrightarrow$	$PH_4^+ + PH_2^{\cdot}$	10.5
	$\longrightarrow$	$P_2H_4^{+\cdot} + H_2$	0.1
	$\longrightarrow$	$P_2H_5^+ + H^{\cdot}$	0.2
$CH_3PH_2^{+\cdot} + CH_3PH_2$	$\longrightarrow$	$CH_3PH_3^+ + CH_4P^{\cdot}$	8.8
	$\longrightarrow$	$CH_3P_2H_4^+ + CH_3^{\cdot}$	0.7

(3.14)   Gas-phase ion-molecule reactions of phosphines

3.4   Electron spectroscopy

Electron spectroscopy will probably, in the next few years, greatly increase our understanding of bonding in phosphorus compounds. At the present time relatively few results are available. X-ray photoelectron spectroscopy has been used to determine binding energies for low-lying stable electrons in various phosphorus compounds (3.15). Because Hartree-Fock calculations of the $2p$ level in isolated P(0) give 140–7 eV and in isolated P(I), 150–6 eV, the data in (3.15) indicate the relatively minor changes that occur in the compounds studied, although substantial C.N. and oxidation state changes are involved. These data are accounted for, qualitatively, by simple electronegativity considerations (with the exception of PH_3 perhaps), and correlations with CNDO molecular orbital calculations of binding energy levels are reasonably good. Although no d-orbital contribution was included in these calculations, it is not yet apparent whether the rather small effects of d orbitals on these low-lying energy levels would be detectable by this technique.

Compound	Binding energy, P_{2p} (eV)	Compound	Binding energy P_{2p} (eV)
H_3P	136.9	Cl_3P	139.6
$(CH_3)_3P$	135.8	Cl_3PO	140.9
$(CH_3)_3PBH_3$	137.0	Cl_3PS	140.4
$(CH_3)_3PCH_2$	137.0	$(C_6H_5)_3P$	129.8
$(CH_3)_3PNH$	137.4	$((C_6H_5)_3P)_2HgI_2$	132.0
$(CH_3)_3PO$	137.6	Na^4PO_4	134.0
$(CH_3)_3PS$	137.4	KPF_6	139.5

(3.15)   Binding energies of the 2p electrons on phosphorus

Photoelectron spectroscopy, in contrast to the X-ray method, determines binding energies for the highest occupied orbitals in compounds studied. For example, in the series $(CF_3)_nP(Cl)_{3-n}$, the ionization energy (or binding energy) of the phosphorus lone pair decreases with increasing chlorine substitution as shown (3.16). Because the inductive effects of a CF_3 group and a Cl atom are very similar, this effect is interpreted as a destabilization of the phosphorus lone pair by the introduction of chlorine lone-pair electrons adjacent to it.

	Compound	P lone-pair binding energy (eV)
	$(CF_3)_3P$	11.70
(3.16)	$(CF_3)_2PCl$	11.13
	CF_3PCl_2	10.70

4 The bond to phosphorus

4.1 Valence bond considerations

4.1.1 Electronic energy levels of phosphorus atoms

Phosphorus has a high first ionization energy (10.48 eV) and a low electron affinity; consequently, the chemistry of phosphorus is overwhelmingly a covalent chemistry. The energy levels of various electronic configurations in isolated atomic phosphorus are shown in (4.1), and the results of recent cal-

$$\text{I.E.} = 1{,}012 \text{ kJ mol}^{-1}$$

Energy	1,000			$\}3s^23p^34d$	
level	800	$\}3s3p^4$	$\left.\begin{array}{l}3s^23p^23d\\3s^23p^24s\end{array}\right\}$	$\}3s^23p^24p$	$\}3s^23p^25s$
above	600				
ground	400				
state	200	$\left.\begin{array}{l}\end{array}\right\}3s^23p^3$			
(kJ mol^{-1})	0				

(4.1)   Energy levels of atomic P

culations of electronic energies of P(0) and P(I) in a number of electronic configurations are given in (4.2). Promotional energies of excited states are

	Configuration	Term symbol	Energy (eV)	Promotional energy from ground state (eV)
P(0)				
	$3s^23p^3$	^4S	−9,248.9	0
	$3s^13p^4$	^4P	−9,240.8	8.1
	$3s^23p^23d^1$	^4F	−9,241.6	7.3
	$3s^13p^33d^1$	^6D	−9,238.2	10.7
P(I)	$3s^23p^2$	^3P	−9,239.9	0 (taken as ground state)

25

	Configuration	Term symbol	Energy (eV)	Promotional energy from ground state (eV)
(P+)	$3s^1 3p^3$	^5S	−9,236.4	3.5
	$3s^2 3p^1 3d^1$	^3F	−9,229.8	10.1
	$3s^1 3p^2 3d^1$	^5F	−9,225.3	14.6

(4.2)   Total electronic energies in $P(0)$, $P(I)$

not extremely large when compared with the magnitudes of typical bond energies. Thus it is hard to rule out, on this basis alone, participation of higher orbitals in the bonding of phosphorus.

4.1.2   Simple hybridization arguments

In phosphorus of C.N.3, because angles at phosphorus are normally not much greater than 90°, a hybridization of close to p^3 is assumed in a valence bond approach. Appropriate hybridizations for C.N.4-6 are the normal ones for the geometries indicated in (4.3).

C.N.	Geometry	Hybridization	Comment
3		p^3	Angles at P 90°
4		sp^3	Tetrahedral
5		$sp^3 d$	Trigonal bipyramidal
6		$sp^3 d^2$	Octahedral

(4.3)   Geometry and hybridization at phosphorus

  Hybridization arguments can be applied successfully to a number of aspects of phosphorus chemistry. For example, the gas-phase basicities and angles at phosphorus for two phosphines are given in (4.4). It can be argued that the increased angle at phosphorus in dimethylphosphine as compared with phos-

Compound (B)	Angle at P (deg)	pKa (g) of BH$^+$
PH$_3$	93.5	-14
(CH$_3$)$_2$PH	99.2	~4.5

(4.4) Phosphine basicities

phine indicates greater s character in the bonding orbitals used by phosphorus. Because the $3s$ orbital on phosphorus is lower lying and more electron attracting than the $3p$ orbital, this would lead to a more electronegative phosphorus atom in dimethylphosphine than in phosphine, and hence would account qualitatively for the increased basicity of the dimethylphosphine. (This argument is incomplete, of course, because the inductive effects of the methyl groups are not discussed explicitly; cf., however, Section 5.2.) Another example of hybridization arguments in the rationalization of PP coupling constants was presented earlier (Section 3.2.1).

A comparison of bond angles in related nitrogen and phosphorus compounds (4.5) provides an interesting test of hybridization arguments. A gen-

Compound	Bond angle (deg)	Compound	Bond angle (deg)
NH$_3$	107	P(CH$_3$)$_3$	99.1 (CPC)
NF$_3$	102.5	PCl$_3$	100
PH$_3$	93.5	PF$_3$	100

(4.5) Bond angles in some related N and P compounds

eral principle that is important here is that s character concentrates in orbitals utilized in binding to more electropositive substituents. (Because s orbitals of the same principal quantum number are lower in energy than p orbitals, and s electrons are therefore more tightly bound, the "looser" p electrons are more easily attracted by the more electronegative ligands, leaving more s character in the bonds to more electropositive ligands.) Increasing s character leads to a larger bond angle (sp^3, 25% s, 109°; sp^2, 33% s, 120°; etc.) and hence we see that the difference in bond angle between NH$_3$ (107°) and NF$_3$ (102°) is rationalized by a simple hybridization argument.

This is not the case, however, for the angles at phosphorus in the compounds presented in (4.5). In fact, in going from PH$_3$ to PF$_3$ (or PCl$_3$), the bond angle increases, and it is clear that the simple hybridization argument is incomplete. We must either invoke ad hoc factors to explain this change (steric factors, mutual repulsion of more charged ligands, multiple bonding to fluorine, etc.) or turn to more sophisticated explanations.

4.2 Results of molecular orbital calculations

There have been some detailed ab initio molecular orbital (MO) calculations on simple molecules containing phosphorus. The results of such a calculation for the highest occupied orbitals of PH_3 are presented in (4.6) and are inter-

| Relative energy (atomic units) from ionization energy | Degeneracy | Orbital constitution | | | | Assignment |
| | | Phosphorus orbitals (%) | | | Hydrogen orbital (%) | |
		3s	3p	3d	1s	
- 0.3683	Single	17	71		4	Lone pair
- 0.5036	Double		47	6	15	Three bonding pairs
- 0.8268	Single	66	3		11	

(4.6) Molecular orbital results for orbital occupancy in PH_3

esting to compare with the results of simple hybridization arguments. In PH_3, because the HPH angle is only a little more than 90° (93.5°), hybridization would assign an almost purely p character to the PH bonding electron-pair orbitals, with the slightly uneasy consequence that the lone-pair electrons would have to be assigned to an orbital of almost pure s character, with no particular directionality. The M.O. results, however, indicate that the HPH bond angle of 93.5° is also accounted for by three bonding orbitals that are not all equivalent energetically, the lowest lying having substantial s character, and the higher doubly degenerate orbital having much p character mixed with a little $3d$ and some hydrogen $1s$ character. The highest occupied orbital, assigned to the lone pair, has mostly p character and is strongly directional. It is salutary to recall that even when bond angles are very close to those expected from simple hybridizations, there may be alternative explanations for their values. The charge distribution calculated for PH_3 by the M.O. method is $\overset{-0.32}{P}\overset{+0.11}{H}_3$, with more negative character at phosphorus than would be expected from consideration of Pauling electronegativities ($X_H = X_P = 2.2$). Molecular orbital calculations for PF_3 also rationalize its bond angle, and give, as expected, much positive character to phosphorus: $\overset{+1.02}{P}\overset{-0.34}{F}_3$.

4.3 Multiple bonding

When the bond to phosphorus is being considered for other than C.N.3, the question of multiple bonding to phosphorus must be discussed. For example, with C.N.4, there are clearly two types of compounds to be examined. If we consider the fourth bond to phosphorus as being formed by a Lewis acid/ Lewis base reaction of a C.N.3 compound (4.7), then the acceptors A can be

(4.7) $R_3P: + A \longrightarrow R_3P{:}A$

Lewis   Lewis
base   acid
        acceptor

classified into two groups. If A can be considered as a pure σ acceptor – for example, H^+ or CH_3^+ – then all four bonds to phosphorus can, as a first approximation, be regarded as σ bonds, and the C.N.4 compound will then be approximately described in terms of sp^3 hybridization at phosphorus.

However, when A is potentially a π donor or acceptor, for instance when A is oxygen, sulfur, a transition metal atom, or a CR_2 group, then there is the possibility of a π bond supplementing the σ bond framework. If A is a π donor, then the π bond may involve acceptance of π electrons into the vacant $3d$ orbitals on phosphorus. Thus we could consider alternative extreme descriptions of the PO bond in a tertiary phosphine oxide, R_3PO. In view of the formal similarity between tertiary phosphine oxides, R_3PO, and tertiary amine oxides, R_3NO, we might adopt a zwitterionic formulation $R_3\overset{+}{P}-\overset{-}{O}$ for the former as we are forced to do for the latter, $R_3\overset{+}{N}-\overset{-}{O}$. (Because nitrogen has no low-lying vacant orbitals that can accept electron density from oxygen via π bonding, the amine oxides are given the zwitterionic formulation, in accord with their large dipole moments and generally polar character.)

By π overlap between oxygen and phosphorus $[(2p_\pi)_O\,(3d_\pi)_P$ for example], however, electron density in phosphine oxides could be partially delocalized with a potential lowering of energy. We can write this in conventional resonance notation (4.8) with the following important reservations.

$$R_3\overset{+}{P}-\overset{-}{O} \longleftrightarrow R_3P{=}O$$

(4.8)   Resonance formulation of phosphoryl group

First, the actual contribution of each form will be strongly dependent on the nature of R. Second, the meaning of the second line in the doubly bonded formulation, $P{=}O$, is that there is overlap between a filled $2p$ orbital on oxygen and a vacant π orbital of suitable symmetry on phosphorus. This π orbital is often taken to be a phosphorus $3d$ orbital, or a combination of $3d$ orbitals, and the resulting π bonding is often referred to as p_π-d_π bonding. It should be pointed out that the properties of p_π-d_π π bonds may be markedly different from those of the more familiar p_π-p_π π bonds found in alkenes, ketones, and so on.

The evidence in favor of a multiple-bond formulation for some bonds involving phosphorus includes a wide range of physical data – thermodynamic, structural, and spectroscopic. The totality of the evidence is quite convincing.

A formulation involving some multiple-bond character is more chemically satisfying, and more useful pedagogically and diagnostically, than a purely zwitterionic description. The evidence will now be examined in detail, because the debate on multiple bonds in third-row and heavier elements continues to be an active area of chemical polemic.

4.3.1 Bond energy data

Thermochemical information available for phosphorus compounds is sketchy and sometimes of dubious quality. There are experimental difficulties in obtaining reliable data; for instance, oxygen combustion of organophosphorus compounds sometimes gives rise to ill-defined phosphorus oxide mixtures.

Although the importance of good thermochemical data for the understanding of chemical reactivity and bonding cannot be stressed too highly, thermochemistry is not regarded by most chemists as a glamorous or attractive area of the science. Hence there are relatively few careful investigations; and this situation is unlikely to change in the near future.

There are, however, enough studies on compounds with potential multiple bonds to phosphorus for some conclusions to be possible. Consider the PO bond in compounds of the R_3PO type. By assuming that the bond energy terms for the RP bonds remain unaltered in going from R_3P to the corresponding R_3PO, one obtains a bond energy term for the PO bond of about 540 kJ mol^{-1} (130 kcal mol^{-1}). In C.N.3 compounds of type $(RO)_3P$, the bond energy term for a model single PO bond is around 380 kJ mol^{-1} (90 kcal mol^{-1}). Thus the bond in the C.N.4 compound is stronger by some 160 kJ mol^{-1} than a model single PO bond. We can assess the likely strengthening by polar character alone of a $\overset{+}{P}-\overset{-}{O}$ bond by turning to the amine oxides as models, though here, too, the thermochemical data are fragmentary and by no means as complete as one would wish. However, in tertiary amine oxides, the bond energy term for the $\overset{+}{N}-\overset{-}{O}$ bond is usually given as 210–90 kJ mol^{-1} (50–70 kcal mol^{-1}), whereas the term for an NO single bond in, for example, hydroxylamine derivatives R_2N-OR', is about 150–90 kJ mol^{-1} (40–50 kcal mol^{-1}). So the bond strengthening from polarity alone in an NO single bond is at most 120 kJ mol^{-1} (30 kcal mol^{-1}), and is probably substantially less than this. Because the polar properties (e.g., bond dipole moment) of a $\overset{}{>}PO$ bond are much lower than those of an $\overset{+}{>N}-\overset{-}{O}$ bond, we could argue that bond-strengthening effects arising from polarity alone in the PO bond would be lower than the maximum noted for the $\overset{+}{N}-\overset{-}{O}$ bond and that, therefore, the 160-kJ mol^{-1} (40-kcal mol^{-1}) strengthening of the $>PO$ bond over a model PO single bond must be attributed, in large part, to multiple bonding. Similar arguments can be made for the PN bond in phosphorus imides; for instance, the bond energy term assigned to the PN bond in $(C_6H_5)_3PN(CH_3)$ is about 420–530 kJ mol^{-1} (100–25 kcal mol^{-1}), whereas that assigned to a model PN single bond in $[(CH_3)_2N]_3P$ is about 290 kJ mol^{-1} (70 kcal mol^{-1}). Here again is a bond strengthening of more than 130 kJ mol^{-1} (30 kcal mol^{-1}), which can be best accounted for in terms of multiple bonding.

4.3.2   Bond length data

The general correlation between bond length and bond order is well established in organic chemistry, especially for carbon–carbon bonds. There have been attempts to place this correlation on a mathematical basis, but we shall treat the correlation as empirical. In general, in carbon chemistry, shortening of a bond below an expected value is taken as implying an increase in bond order, that is, the occurrence of some multiple bonding. However, development of ionic character in a bond is not normally expected to produce marked changes in bond length, because as the ionic radius of the cationic center diminishes so that of the anionic center increases, and consequently the two effects tend to cancel. For example, the sum of the covalent radii for singly bonded nitrogen (r = 70 pm, 0.70 Å) and oxygen (r = 66 pm, 0.66 Å) is 136 pm (1.36 Å), and a wide variety of NO single bonds have lengths quite close to this, whether the atoms carry a formal charge or not (4.9). Thus we shall

Compound	r_{NO} (pm)
$H_2NC(H){:}NOH$	141
$(CH_3)_3\overset{+}{N}\overset{-}{O}$	140
$[(CH_3)_3\overset{+}{N}OH]\overset{-}{Cl}$	142

(4.9)   Some NO single bond lengths

take a marked bond length contraction as evidence supporting multiple bonding (though not, of course, as proving it). The data for PC, PN, and PO multiple bonds are given in (4.10). Some comments on the defining compounds are in order.

First we note that there are major changes in bond length with compound type for all three bonds, suggesting the occurrence of true multiple bonding. For the CP bonds the single bond is represented by $(CH_3)_3P$. In the gas phase the high-temperature species CP and the molecule HCP both have the same CP bond length. Spectroscopic evidence for HCP, particularly ^{13}C nmr in solution, which shows the CH bond to be *sp* hybridized, confirm the existence of a CP triple bond in this compound. With these end points it is gratifying to find that the crystalline phosphorus ylide $(C_6H_5)_3PCH_2$ has a PCH$_2$ bond length that a PC double bond would be predicted to have; and that the ring PC bonds in the substituted phosphabenzene shown in (4.10) are of a length corresponding to a PC bond order of 1.5.

The PN bond correlation plot has fewer points, due in large measure to

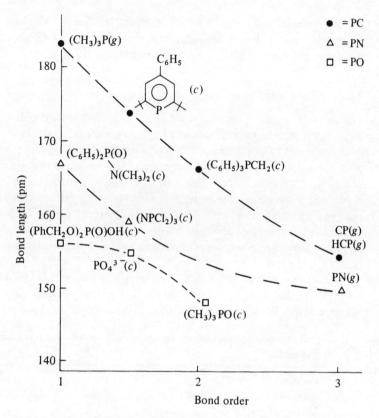

(4.10) Bond order/bond length plots for PC, PN, and PO bonds

the relative paucity of experimental data. Even a good model compound for the PN single bond is elusive; it could be argued that the one chosen, $(C_6H_5)_2P(O)N(CH_3)_2$, has some multiple-bond character in the PN bond due to a resonance interaction:

The triple PN bond length, oddly enough, is less in contention, since $PN(g)$ is a well-established species. With these end points the PN bond length in the cyclophosphazenes falls nicely at a PN bond order of 1.5.

The PO correlation plot is the least well established of the three, and is still uncertain in some details. Again, few good models are available. Although the choice of single- and double-bond models seems reasonable, the 1.5 bond order in PO_4^{3-} may be debatable (see below).

4.3.3   Infrared spectroscopic data

As previously discussed (Section 3.1), the P=O bond is a strong absorber in the infrared, and the absorption frequency is a sensitive function of the electronegativity of the other groups attached to phosphorus. A Hooke's law model for the P=O bond vibration leads to a relationship $K_{PO} \alpha \sqrt{\nu_{PO}}$, where K is the stretching force constant for the bond and ν is the absorption frequency. This model assumes, of course, that the observed absorption is a pure PO stretching frequency. The stretching force constant K is, in its turn, proportional to bond multiplicity in the simplest model. Hence, if the various assumptions implicit in the above arguments hold, then from ν_{PO} one can deduce both K_{PO} and the order of the PO bond. Typical results are presented in (4.11), and they are in reasonable accord with conclusions drawn from bond length arguments.

Compound	K_{PO} (Nm^{-1})	PO bond order
PO_4^{3-}	630	1.5
$(CH_3)_3PO$	900	1.8
Cl_3PO	1,040	1.95
$(CF_3)_3PO$	1,110	2.0
P_4O_{10} (exocyclic PO bonds)	1,100	2.0
F_3PO	1,240	2.0

(4.11)   PO bond orders from vibrational frequencies

4.3.4   Orbital involvement in multiple bonding

The exact nature of the orbitals involved in the multiple bonds we have been discussing has been debated by theoretical chemists for some time. It was Pauling in 1931 who first suggested d-orbital involvement in phosphorus bonding, to explain the existence and certain properties of PF_5 and PF_6^- and, as Coulson later commented, theoretical chemists were for a long time so bemused by the qualitative beauty of this idea that they failed to put it to the quantitative test. Although in the neutral phosphorus atom the $3d$ orbital (r_{max} = 243 pm, 2.43 Å) is far too diffuse to be involved in bonding, it is now clear that Craig's qualitative suggestion, in 1954, that electronegative substituents may contract the $3d$ orbitals on phosphorus, making them more suitable for involvement in covalent bonds, can be supported quantitatively. Thus the general picture we draw of these multiple bonds is one involving donation of electrons from, say, the filled oxygen $2p$ orbitals into a vacant $3d$ orbital on phosphorus, yielding a multiple $p_\pi-d_\pi$ π bond that is relatively polar, because the region of maximum overlap is closer to oxygen than to phosphorus (4.12).

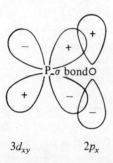

$3d_{xy}$ $2p_x$

Molecular orbital calculations support this picture. Thus, in considering the change $(CH_3)_3P: + O \longrightarrow (CH_3)_3PO$, the new PO bond involves σ donation from phosphorus to oxygen and π back-donation from oxygen to phosphorus. The calculated charge on phosphorus in the resulting oxide is $+0.35$, which is almost the same as the charge on phosphorus in the original phosphine! Similar observations can be made for the F_3PO molecule (4.13). Again the charge

$$\begin{array}{ccccccc} -0.34 & +1.02 & & & -0.32 & +1.11 & -0.15 \\ (F)_3 & P & +O & \longrightarrow & (F)_3 & P & O \end{array}$$

(4.13)   MO results for charge distribution in formation of PO bond

on phosphorus scarcely changes during PO bond formation. Thus the MO picture of formation of this bond corresponds more closely to a somewhat polar multiple bond than to a highly polar (zwitteronic) single bond.

4.3.5   Cyclophosphazenes, a special class of multiply bonded systems

The cyclophosphazenes, $(NPX_2)_n$, can be prepared by various reactions (4.14)

$$NH_4Cl + PCl_5 \xrightarrow[C_2H_2Cl_4]{reflux} (NPCl_2)_n \xrightarrow{SbF_3} (NPF_2)_n$$
$$\begin{array}{cc} n = 3, 4, 5, 6, 7; & n = 3, 4, \ldots, 17 \\ \text{plus higher polymers} & \text{characterized} \end{array}$$

$$(CF_3)_2PCl + NaN_3 \longrightarrow [(CF_3)_2PN_3] \xrightarrow[-N^2]{\Delta} [NP(CF_3)_2]_n$$

(4.14)   Preparations of cyclophosphazenes

and have a long history and an elaborate chemistry. Many different nucleophiles attack these compounds at phosphorus, yielding substitution products, and the derivative chemistry of the rings, their stereochemical aspects, and many structural questions have provided chemists with years of worthwhile work – and continue to do so. In this section we confine ourselves to a discussion of bonding in the ring system.

The trimeric $(NPCl_2)_3$ has a planar N_3P_3 skeleton with all PN bond lengths equal (r_{PN} = 159 pm, 1.59 Å) and of a length compatible with bond order 1.5. The tetrameric $(NPCl_2)_4$ is not planar, but the physical and chemical properties of trimer and tetramer and, indeed, of the larger ring compounds also, are very similar. Thus all of the cyclophosphazenes have similar uv spectra, with $\lambda_{max} \approx 200$ nm, and with $\epsilon_{max} \alpha n$, the number of phosphazene units in the ring. The little thermochemical data available indicate that the strength of the PN bond hardly changes with ring size.

These properties are not compatible with any bonding picture that implies more and more extensive conjugation between PN systems as ring size increases. Nor are they compatible with a unique "aromatic" character for the trimer (4.15) as opposed to a "polyene" character for the larger ring com-

(4.15)   An implausible resonance scheme

pounds. The best qualitative description of the cyclophosphazenes is that of Dewar (4.16), who suggested that adjacent PNP units could be treated as iso-

(4.16)   PNP π MOs in cyclophosphazenes

lated conjugated three-centered "island" π orbitals (somewhat akin, electronically, to those of an allyl cation), which are insulated from each other by a node at each of the terminal phosphorus atoms. In this way the cyclophosphazene is seen to be nonaromatic, because the π system is not fully delocalized over the whole molecule, but it does have a large resonance energy (ca. 0.8 β per P_2N unit), and the constancy of the uv λ_{max} and PN bond energy with increasing ring size are readily understood.

4.3.6 P_π-P_π multiple bonds

In compounds with lower coordination number, such as C.N.1 or C.N.2, phosphorus forms p_π-p_π multiple bonds with carbon or nitrogen. These compounds have already been briefly described (Sections 2.1 and 2.2), and in this section we shall be concerned with a description of the phosphorus bonding in them. A justification for a triple bond in HC≡P, involving $2p_\pi$-$3p_\pi$ carbon-phosphorus overlap was presented above (Section 4.3.2) and was based on bond length data, and on interpretation of nmr observations of the H-^{13}C coupling constant (211 Hz) as being consistent with sp hybridization at carbon.

In the phosphabenzenes, bond length data is again consistent with a π interaction of the phosphorus $3p_\pi$ orbitals with carbon $2p_\pi$ orbitals (Section 4.3.2). In addition the intense uv spectra of the phosphabenzenes, the frequencies of their ring vibrations, and the low-field proton nmr shifts of the ring protons (4.17) are all consistent with an aromatic delocalization involving $3p_\pi$-$2p_\pi$ bonding.

λ_{max} 213 nm (ϵ, 19,000)
246 nm (ϵ, 8,500)

Ring modes, 1,515, 1,395 cm^{-1}

Planar (microwave spectrum, gas)

$^\delta H\alpha$ 8.9 $^\delta H\beta$ 7.5 $^\delta H\gamma$ 8.0

(4.17)   Some physical properties of phosphabenzene

5   Phosphorus of C.N.3

5.1   Notes on synthesis

The primary sources of C.N.3 phophorus compounds in industry are white
phosphorus, P_4, prepared by thermal carbon reduction of phosphate rock,
and phosphorus trichloride, PCl_3, produced by controlled chlorination of
P_4. Because of the growing interest in phosphorus chemistry, many other
three-coordinate derivatives are now commercially available in quantity.

Upon fusion with metals, elemental phosphorus (usually the safer amorphous red allotrope is used) gives metal phosphides; hydrolysis of these gives
phosphine, PH_3, and, in some cases, moderate amounts of the thermally
unstable diphosphane, P_2H_4. Organophosphines can be prepared from phosphine either by addition reactions, which can be free radical or base catalyzed,
or by nucleophilic substitution involving phosphinide anions. The scheme in
(5.1) shows examples of the synthetic utility of phosphine.

(5.1)   Phosphine in synthesis

Phosphorus trichloride, PCl_3, easily undergoes controllable nucleophilic
substitution reactions at phosphorus with alkoxide ions, primary and secondary amines, mercaptide ions, halide ion sources, and so on. Reactions with
organometallic agents are usually less controllable and generally lead to
tertiary phosphines. Phosphorus trichloride also can be used in aromatic

substitution reactions and, via complexes with $AlCl_3$, can be partially substituted with aliphatic groups. The scheme in (5.2) gives examples of these and related reactions.

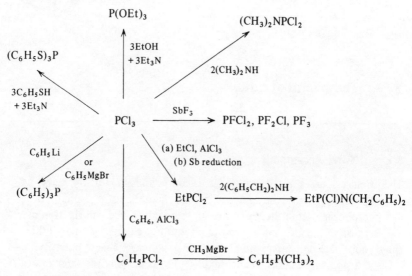

(5.2) Some syntheses from PCl_3

5.2 Stereochemical aspects of C.N.3

5.2.1 Pyramidal structure

Three-coordinate phosphorus compounds [with the so-far unique exception noted in (Section 2.3)] have trigonal pyramidal structures and, in contrast to their nitrogen analogs, have rather high barriers to thermal inversion. Consequently, resolution of phosphines into robust enantiomers is possible in principle, though in practice most optically active phosphines have been prepared by stereoselective or stereospecific processes starting from optically active C.N.4 precursors. Some typical routes to optically active phosphines are shown in (5.3).

$$(CH_3)(nC_3H_7)\overset{+}{P}(C_6H_5)(CH_2C_6H_5) \xrightarrow[\text{at Hg}]{2e^-, H^+}$$
$(+)\alpha D$; resolved via diastereomeric salt $\quad$ cathode

$$PhCH_3 + (CH_3)(nC_3H_7)P(C_6H_5)$$
$$[\alpha]_D = +16.8°$$

$$(CH_3)(nC_4H_9)\overset{+}{P}(CH_2C_6H_5)(CH_2CH_2CN) \xrightarrow[\text{MeO}^-]{\text{MeOH}}$$
$(+)\alpha D$; resolved via diastereomeric salt

$$CH_2{=}CHCN + (CH_3)(nC_4H_9)P(CH_2C_6H_5)$$
$$\text{optically active}$$

$$\underset{\substack{\text{resolved via crystallization}\\ \text{of diastereomers}}}{R^1 R^2 \overset{\displaystyle O}{\overset{\|}{P}}[-O(-)\text{menthyl}]} \xrightarrow{\;R^3 MgBr\;}$$

$$\underset{\text{active}}{R^1 R^2 R^3 PO} \xrightarrow{\;Si_2 Cl_6\;} Si_2 Cl_6 O + \underset{\substack{\text{optically}\\ \text{active}}}{R^1 R^2 R^3 P}$$

(5.3)   Some preparations of optically active phosphines

5.2.2   *Barriers to inversion*

There has been much interest in the determination of barriers to inversion in phosphines; the principal methods used have been polarimetric studies of optically active phosphines or nmr studies of compounds that have diastereo-topic groups. In some of the nmr studies, only lower limits to inversion barriers could be established. The inversion process is of considerable general interest in chemistry, because it is an example of a first-order unimolecular process in which the nature of the transition state (involving a planar con-figuration around phosphorus) is well defined (5.4).

planar transition
state

(5.4)   Thermal inversion at phosphorus

The three principal effects on inversion barriers at phosphorus of C.N.3 are (1) steric effects, (2) electronegativity effects, and (3) effects of conjugation. Each of these will now be examined briefly.

Steric effects. Although only a few results are available (there are many more available for nitrogen C.N.3 compounds), it does seem as if increasing the bulk of substituents at phosphorus decreases the inversion barrier [see (5.5) for examples). This is generally rationalized by invoking a ground-state flat-tening of the pyramid at phosphorus as bulky substituents are introduced, thus making the planar transition state more easily reached. It would be of considerable interest to check this interpretation by structure determinations, or even by the easier examination of other ground-state properties such as ^{31}P chemical shifts or ^{13}C-^{31}P coupling constants, but this has so far not been done.

R $\Delta H^{\ddagger}$ (kJ mol^{-1})

CH$_3$ $\geqslant 145$ (could not be estimated more precisely)

(CH$_3$)$_3$C 117 ± 4

(5.5)   Steric effects on inversion barrier at P

Electronegativity effects. As the electronegativity of substituents at phosphorus increases, so does the inversion barrier. In fact, a plot of $\Delta G^{\ddagger}$ for inversion versus the sum of the Allred-Rochow electronegativities of substituent groups at phosphorus is accurately linear for some 20 examples (both observed barriers and barriers calculated by CNDO molecular orbital methods). An exemplary series is presented in (5.6). The rationalization of this effect by a hybridization argument is as follows. As substituent electronegativity increases, the hybridization of the orbital bonding to the substituent gains in p character in the ground state (cf. Section 4.1), leaving the phosphorus lone pair in an orbital of more s character in the ground state. In the planar transition state, however, the phosphorus lone pair must be in an orbital of pure p character. Hence the energy involved in the rehybridization necessary to achieve the transition state is greater, the more electronegative the substituents at phosphorus. Although the general correlation of inversion barriers with electronegativity seems to be well founded, there have been criticisms of the detailed application of the particular electronegativity scale (Allred-Rochow) that gives the best quantitative correlations, and the debate on this issue is not yet closed.

M	R	$\Delta G^{\ddagger}$ (kJ mol^{-1})	X_M (Allred-Rochow)
C	CH$_3$	137	2.50
Si	(CH$_3$)$_2$CH	79	1.74
Ge	(CH$_3$)$_2$CH	89	2.02
Sn	(CH$_3$)$_2$CH	81	1.72

(5.6)   Free energy of activation for inversion at P, correlated with electronegativity

Conjugation effects. Conjugation of the phosphorus lone-pair electrons with an adjacent unsaturated center, either in the ground state or in the transition state, can lead to dramatic reductions in inversion barriers. Examples for acylphosphines and for phospholes (phosphacyclopentadienes) are given in

(5.7). The infrared data given in (5.7) suggest that conjugation may be important in the ground state in acylphosphines, thus flattening the phosphorus pyramid, but no structural data are available. In the phospholes, conjugation does not seem important in the ground state. Phospholes are not flattened at phosphorus, and behave chemically as typical dienes in, for example, Diels-Alder reactions, with no indication of aromaticity. However, conjugation, in the planar transition state, of the lone-pair electrons, which are then in a pure p orbital, with the unsaturated system presumably leads to an "aromatic" transition state, which is thus stabilized by some 80 kJ mol^{-1} with respect to a saturated analog.

Compound	$\Delta G\ddagger$ for inversion (kJ mol^{-1})
	152
	67
$C_6H_5P[CH(CH_3)_2](CH_2CH_3)$	134
$C_6H_5P[CH(CH_3)_2](COCH_3)$	81
Compound	$\nu_{CO}(cm^{-1})$
$C_6H_5COCH_3$	1,686
$C_6H_5CONH_2$	1,652
$C_6H_5COP(nC_5H_{11})_2$	1,652

(5.7)   Inversion barriers and spectral data in conjugated systems

5.2.3   Barriers to rotation

A rather unexpected aspect of the stereochemistry of C.N.3 phosphorus compounds was the observation of substantial barriers to rotation about certain formally single PN bonds. Some examples are given in (5.8).

System	$\Delta G\ddagger$ (kJ mol^{-1} for PN rotation)
$CH_3P(Cl)N(CH_3)_2$	48.1
$CH_3P(Cl)N[CH(CH_3)]_2$	56.0
$CH_3P(F)N[CH(CH_3)_2]_2$	45.6

(5.8)   Some PN rotation barriers

These barriers are, of course, substantially higher than those normally associated with single-bond rotation. It must be noted, however, that in terms of the wider context of the whole periodic table, our knowledge of the energetics of bond rotation is fragmentary, with most of the available data coming from carbon chemistry. In extensions of the work on C.N.3 phosphorus compounds, substantial barriers have been detected in C.N.4 and C.N.5 compounds of phosphorus, and for analogous silicon-, phosphorus-, and sulfur-nitrogen compounds, the barriers increase in the order SiN $<$ PN $<$ SN. There is evidence that the major interaction giving rise to the rotational barrier is a lone-pair/lone-pair interaction, but there is also the possiblity of some phosphorus-nitrogen π interaction, with an angular dependence. This latter receives some support from the observation that in a number of C.N.3 P-N compounds, the geometry at nitrogen is trigonal planar.

5.3 Phosphorus as a base and nucleophile

5.3.1 Phosphine basicity

In aqueous solution most phosphines are rather weak bases, and so the pK_a values of their conjugate acids have been established mainly by nonaqueous titrations, for instance, potentiometrically in nitromethane solution, with nitrogenous bases providing standards. There are not many values available, but for a number of tertiary phosphines, R_3P, and secondary phosphines, R_2PH, these solution pK_a values correlate well with the Taft σ^* values of the substituents at phosphorus (5.9); σ^* values measure the pure inductive effects

$$\text{For tertiary phosphines, } pK_a = 7.85 - 2.67 \ \Sigma\sigma^*$$
$$\text{For secondary phosphines, } pK_a = 5.13 - 2.61 \ \Sigma\sigma^*$$

(5.9) Correlations between pK_a and substituent constants

of substituents. However, for some bulkier groups on phosphorus, such as $(CH_3)_2CHCH_2$, there are deviations from the correlations, and it is clear that a wider and more critical test of this simple inductive effect relationship would be desirable.

Recently, via ion-cyclotron resonance (cf. Section 3.3), gas-phase basicity data have become available for some simple phosphines and can be compared with solution information. A comparison with similar nitrogenous bases is also informative. Some pertinent data are given in (5.10). The cyclohexyl compounds are included as model compounds for the primary and secondary methylphosphines for which no direct enthalpy of solution in fluorosulfonic acid could be obtained.

In the gas phase the proton affinity of phosphine increases steadily with increasing alkylation (as is the case, in the gas phase, for the analogous amines). Replacing a hydrogen on phosphorus by a more polarizable alkyl

Compound (B)	Proton affinity (g) (kJ mol^{-1})	$-\Delta H$ for solution in HSO$_3$F (kJ mol^{-1})	pK_a of BH$^+$
PH$_3$	786	60	ca. -14
CH$_3$PH$_2$	865	–	–
(CH$_3$)$_2$PH	914	–	3.91
(CH$_3$)$_3$P	954	186	8.65
(c-C$_6$H$_{11}$)PH$_2$	–	126	0.27
(c-C$_6$H$_{11}$)$_2$PH	–	136	4.55
NH$_3$	866	181	9.24
CH$_3$NH$_2$	905	193	10.65
(CH$_3$)$_2$NH	931	200	10.79
(CH$_3$)$_3$N	948	198	9.80

(5.10)　Measures of basicity for phosphines and amines

group stabilizes the phosphonium cation. The solution enthalpy results are also quite regular and parallel the pK_a values. Because phosphonium cations $\geq\overset{+}{P}$—H, like other phosphorus-hydrogen compounds, do not appear to hydrogen bond in solution, replacement of H by alkyl in the cation does not destroy hydrogen bonds, but does lead to charge stabilization. However, the strong hydrogen bonding in $\geq\overset{+}{N}$—H compounds leads to a trade-off between hydrogen-bonding capability and alkyl stabilization as H is replaced by alkyl groups in ammonium cations. Hence the maximum in pK_a value for the (CH$_3$)$_2\overset{+}{N}$H$_2$ cation in aqueous solution, as contrasted with the smooth increase in proton affinity in the gas phase in going from ammonia to trimethylamine.

Two further interesting questions arise in examining (5.10). Why is phosphine so much less basic in the gas phase than ammonia? And why has the gas phase basicity difference virtually vanished when we examine trimethylphosphine and trimethylamine? The best available answer to the first question seems to lie in the rehybridization energy needed in the conversion of PH$_3$ (with its lone-pair electrons in an orbital of character describable as $sp^{0.8}$) to PH$_4^+$, in which all bonds are of the sp^3 type. In ammonia, in contrast, there is no rehybridization energy needed, because the bonds (and lone pair) in NH$_3$ are already sp^3 in character, as they are in NH$_4^+$ (see Sections 4.1 and 4.2).

The extra effectiveness of methyl stabilization of phosphonium cations, as opposed to ammonium cations, is less easy to explain. It may be due to an increasing degree of rehybridization needed for the amines with increasing methyl substitution, for which there is marginal evidence from dipole moment studies, but at present it must be said that even such an apparently simple phenomenon as the effect of an alkyl group on phosphine basicity is not fully explicable.

Another way of estimating gas-phase basicities of 3-C.N. phosphorus compounds that has been recently proposed is the use of phosphorus lone-pair ionization energies, which can be measured by photoelectron spectroscopy (Section 3.4). If one considers the enthalpy changes in the heterolytic dissociation, in the gas phase, of acid–base adducts as a good measure of base strengths toward a particular acid, then the lone-pair ionization energies of the bases can be simply related to the required enthalpy changes as shown in (5.11). If A is constant, then, for a series of bases, $S_B = -IP + D_{(PA)}$. There

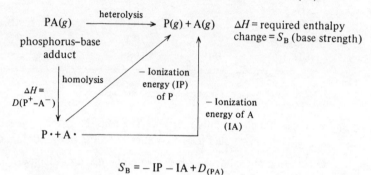

$$S_B = -IP - IA + D_{(PA)}$$

(5.11)   Thermodynamic cycle for base strength estimation

is some evidence that for the particular case where A is the proton, $D_{(PA)}$ remains nearly constant, in which case S_B can be correlated linearly with IP. In (5.12) are listed the lone-pair ionization energies of a number of C.N.3

PH_3	10.58 (786)	PCl_3	10.7	PF_3	12.3
CH_3PH_2	9.72 (865)	CH_3PCl_2	9.83	CH_3PF_2	10.34
$(CH_3)_2PH$	9.08 (914)	$(CH_3)_2PCl$	9.19	$(CH_3)_2PF$	9.37
$(CH_3)_3P$	8.63 (954)				

(5.12)   IP values (eV) and proton affinities (in parentheses and in kJ mol^{-1}) for some C.N.3 compounds.

phosphorus compounds, together with their gas-phase proton affinities where these are known. It can be seen that there is an excellent correlation between the values where they are both available, and hence it seems as if ionization potentials may be a fair measure of basicity toward an uncomplicated acid, such as the gas-phase proton. It is interesting that PH_3 and PCl_3 have such similar IP values, indicating that the inductive effect by electron withdrawal of chlorine is almost balanced by mesomeric electron release. For fluorine, however, inductive electron withdrawal predominates.

5.3.2 Phosphorus nucleophiles

Phosphines and other C.N.3 phosphorus compounds are good nucleophiles toward a variety of electrophiles. Toward carbon electrophiles such as alkyl halides, tertiary phosphines are more powerfully nucleophilic than analogous tertiary amines (5.13). The excellent nucleophilicity of phosphines toward

Compound	$pK_a(aq)$	Relative rate of reaction with C_2H_5I in acetone
$C_6H_5N(C_2H_5)_2$	7	1
$C_6H_5P(C_2H_5)_2$	6	520

(5.13)   Relative nucleophilicities of an amine and a phosphine

carbon is due to their relatively low ionization energies (favoring electron transfer); the relatively good strength of carbon–phosphorus bonds; and the large size and high polarizability of the phosphorus atom, leading to a lower repulsion energy during the reaction.

A number of kinetic studies of the quaternization reactions of phosphorus nucleophiles have been undertaken. The reactions show a generally close correlation between basicity (toward the aqueous proton) and carbon nucleophilicity. The quaternization reaction is generally second order, and its rate is markedly sensitive to steric effects both at phosphorus and at the carbon center under attack (5.14). In a Hammett study of the quaternization reactions of phosphines p-X—$C_6H_4P(C_2H_5)_2$ with ethyl iodide in acetone a negative ρ value was found, as one would expect: $\rho = -1.0$, indicating that

Phosphine	pK_a	Relative rate of reaction with C_2H_5I in acetone
$(CH_3)_2PC_2H_5$	8.62	1
$CH_3P(C_2H_5)_2$	8.62	0.53
$P(C_2H_5)_3$	8.62	0.19

Alkyl halide	Relative rate of reaction with $(nC_4H_9)_3P$ in acetone
CH_3I	1.0
C_2H_5I	5.7×10^{-3}
$CH_3CH_2CH_2I$	2.4×10^{-3}
$CH_3CHICH_2CH_3$	1.9×10^{-4}

(5.14)   Steric effects in phosphine quaternization reactions

electron-withdrawing substituents reduced the reaction rate. For the comparable amine quaternization reactions, however, a much larger negative ρ is observed: $\rho = -2.8$, showing that conjugative effects through the aromatic ring to nitrogen are much more important, much more effectively transmitted, than they are to phosphorus. This must reflect the poorer overlap of the phosphorus lone-pair orbital (principal quantum number 3) than the nitrogen lone-pair orbital (principal quantum number 2) with the ring π orbitals (also of principal quantum number 2), (5.15).

much less important for E = P
than for E = N

(5.15) Conjugation with benzene ring

5.4   Phosphorus ligands in coordination chemistry

Phosphines and other phosphorus compounds of C.N.3 are among the most widely used ligands in coordination chemistry. Coordination can be viewed as an acid/base reaction in the wider Lewis classification, and the phosphorus Lewis bases belong to the "soft" base group of Pearson's hard and soft acid/base (HSAB) theory. This implies that phosphorus bases bind more strongly to "soft" Lewis acids, such as transition metal atoms in low oxidation states. In (5.16) the soft character of a phosphorus base, as shown by its stronger

Base (B)	$pK_h{}^a$	$pK_{Hg}{}^b$
H_3N	9.42	7.60
$(C_2H_5)_3P$	8.8	15.0

$^aK_h = [HB^+]/[H^+][B]$. $^bK_{Hg} = [CH_3HgB^+]/[CH_3Hg^+][B]$.

(5.16) Relative base strengths to $H^+(aq)$ and $CH_3Hg^+(aq)$

binding to the soft methylmercury cation, is contrasted quantitatively with the harder character of an analogous nitrogen base, as shown by its stronger binding to the hard aqueous proton. HSAB is useful for the qualitative explanation of a number of themes in coordination chemistry, for example, the frequent use of phosphorus donors to stabilize the very soft zero-valent states of transition metals, as in the complexes $[(C_6H_5)_3P]_3Pt$, and $(F_3P)_4Ni$. The soft character of phosphorus donors is due to the same combination of properties as was used to explain the high nucleophilicity of phosphorus nucleophiles toward carbon (Section 5.3), namely, low ionization potentials, allowing for easy electron transfer, and high polarizability, allowing for good accommodation of the developing charge. There is also the possibility of one more controversial aspect, namely, the potential ability of the phosphorus ligand to act not only as a σ donor, but also as a π acceptor, and this aspect of metal–phosphorus bonding will be discussed below.

In terms of ligand field concepts, phosphorus C.N.3 ligands are of high ligand field strength, coming close to CN^- and CO in the spectrochemical series, and frequently phosphines or other phosphorus donors can replace CO in metal carbonyls. Thus in the series of substituted chromium hexacarbonyls, all the compounds $Cr(CO)_{6-n}(PR_3)n$, where $n = 0, 1, 2$, and 3, can be prepared. For Fe(0) and Ni(0), all the compounds $Fe(CO)_{5-n}(PF_3)_n$, where $n = 0, 1, 2, 3, 4$, and 5, and $Ni(CO)_{4-n}(PF_3)_n$, where $n = 0, 1, 2, 3$, and 4, are known. The high ligand field strength and good reducing properties (cf. Section 5.6) of many 3-C.N. phosphorus compounds lead to the use of phosphines in the stabilization of hydrido- and organometallic complexes of low-oxidation-state metals. Some examples are given in (5.17).

$$H_3Re[P(C_6H_5)_3]_4$$

$$H\quad P(C_2H_5)_3$$
$$\diagdown Pd \diagup$$
$$Br\quad P(C_2H_5)_3$$

$$(C_2H_4)Pt[P(C_6H_5)_3]_2$$

(5.17) Some low-oxidation-state phosphine complexes

5.4.1 *Phosphorus trifluoride as a ligand*

Phosphorus trifluoride is unusual among 3-C.N. phosphorus donors in its remarkable similarity in donor properties to carbon monoxide. As cited above (Section 5.4), PF_3 and CO are often interchangeable in transition metal chemistry. Like CO, PF_3 reacts with salts and even with some metals to form complexes directly (5.18). The complexes are often volatile (like metal carbonyls) and remarkably robust; $Ni(PF_3)_4$ can be steam distilled without decomposition.

$$Ni_{(powder)} \xrightarrow[70^\circ C]{PF_3} Ni(PF_3)_4$$

$$PtCl_2(c) \xrightarrow[75^\circ C]{PF_3} cis\text{-}PtCl_2(PF_3)_2$$

(5.18) Direct synthesis of PF_3 complexes

The similarity in ligand properties of PF_3 and CO suggests that π-acceptor character may be important in the phosphorus ligand, as it is well known to be in carbon monoxide. Of all phosphorus ligands, PF_3 is the one for which phosphorus vacant d-orbital interaction with filled metal d orbitals is most plausible. The high electronegativity of the fluorine atoms would be most effective in contracting the diffuse phosphorus $3d$ orbitals, leading to more effective overlap with the metal d orbitals. There is some evidence to support this hypothesis.

In the complex $Ni(PF_3)_4$, which has a tetrahedral NiP_4 skeleton, electron diffraction has determined $r_{PF} = 156$ pm (1.56 Å), very close to that in PF_3 (where $r_{PF} = 154$ pm, 1.54 Å), and $r_{NiP} = 210$ pm (2.10 Å). This is markedly shorter than the metal–phosphorus bond length in many metal–phosphine complexes, which is more normally in the range of 230–40 pm, suggesting more multiple-bond character in the metal–phosphorus bond in this PF_3 complex than in complexes of alkyl or aryl phosphines.

There is also some evidence from infrared spectroscopy. As the data in (5.19) show, as the phosphorus ligand changes from PF_3 to PCl_3 to

Compound	$\nu_{CO}(cm^{-1})$
$(F_3P)_3Mo(CO)_3$	2,090, 2,055
$(Cl_3P)_3Mo(CO)_3$	2,040, 1,991
$[(C_2H_5)_3P]_3Mo(CO)_3$	1,937, 1,841

(5.19) ν_{CO} in some P-substituted carbonyls

$P(C_2H_5)_3$, the CO stretching frequency in the substituted molybdenum carbonyls falls, implying a drop in the CO bond order, hence an increase in the metal–carbon bond order, and consequently a decrease in the metal–phosphorus bond order. So this ir data may be interpreted as meaning that the order of decreasing metal–phosphorus bond strength is $PF_3 > PCl_3 > P(C_2H_5)_3$. Of course, whether this increased bond strength in the metal-PF_3 bond is due to a π-bonding component is not settled by this evidence. It is true that the base strengths of these C.N.3 compounds is in the order $(C_2H_5)_3P > PF_3$ toward Lewis acids of harder type, such as H^+ or BH_3, where σ donation would be expected to be the dominant effect. But although the weight of the evidence does lean toward a degree of π back-bonding in the bonds formed by PF_3 and metals, the case is by no means completely proven.

5.4.2 The metal–phosphorus bond in complexes

If there is still some question about the degree of π bonding in the metal complexes of PF_3, the ligand for which π effects are expected to be at their strongest and most obvious, it is not surprising that there is even more debate over the nature of other metal–phosphorus bonds. The most fashionable research technique for examining this problem is metal–phosphorus coupling constants determined by nmr. In evaluating these the Fermi contact term, which depends on s-electron density at the nuclei that are directly coupled, is taken to be the dominant one, and so arguments involving the participation of d orbitals tend to be very indirect, and connected to the actual observations by a rather long chain of reasoning. It is instructive to review sample arguments on both sides of the question.

An argument against π bonding in the Pt–P bond. Consider the $^1J_{PtP}$ values collected in (5.20). If the Fermi contact term is dominant, then we might expect the ratio of $^1J_{PtP}$ for *cis*-PtCl$_2$(PBu$_3$)$_2$, where Pt is dsp^2 hybridized, and therefore the orbital binding to phosphorus has $\frac{1}{4}s$ character; and of $^1J_{PtP}$ for cis-PtCl$_4$(PBu$_3$)$_2$, where Pt is d^2sp^3 hybridized, and the orbital binding to phosphorus has $\frac{1}{6}s$ character, to be $(\frac{1}{4}/\frac{1}{6}) = 1.5$. The actual ratio observed = 3,508 Hz/2,070 Hz = 1.69; the agreement is fair. As the values in (5.20) indicate, the s-orbital character in the Pt–P bond is consistently larger in cis

Compound	$^1J_{PtP}$ (Hz)	
cis-PtCl$_2$(PBu$_3$)$_2$	3,508	
trans-PtCl$_2$(PBu$_3$)$_2$	2,380	Ratio 1.46 for Pt(II)
cis-PtCl$_4$(PBu$_3$)$_2$	2,070	
trans-PtCl$_4$(PBu$_3$)$_2$	1,462	Ratio 1.42 for Pt(IV)

(5.20) Some $^1J_{PtP}$ values

than in trans complexes of similar geometry, and the ratio of s character in the Pt–P bonding orbital in the cis and trans complexes is remarkably similar for Pt(II) (1.46) and for Pt(IV) (1.42). However, π bonding involving donation of metal d electrons into vacant d orbitals on phosphorus should be more important for Pt(II) than for Pt(IV). First, there are more d electrons available in Pt(II) (d^8) than in Pt(IV) (d^6). Second, the higher charge on Pt(IV) makes electron back-donation less favorable.

Additionally, π bonding should be more important for cis complexes than for trans complexes, because in cis complexes two different filled d orbitals on Pt could be involved in overlap with vacant phosphorus d orbitals; in trans complexes only one Pt d orbital can be involved.

However, because π effects are expected to be relatively insignificant in Pt(IV) complexes, for the reasons given above, and because the coupling constant ratio for the cis/trans complexes is virtually the same for Pt(IV) and for Pt(II), the s character effects that we must invoke to explain the Pt(IV) values can be extended to the Pt(II) values, and there is no need to use π bonding to explain any of the observations.

An argument for π bonding in the Pt-P bond. In (5.21) the reduced coupling constants for analogous ^{31}P and ^{15}N complexes of Pt are presented (reduced

Compound[a]	K_{PtN}[b]	K_{PtP}	K_{PtP}/K_{PtN}
trans-PtCl$_2$L$_2$	112.5	228.6	2.03
trans-PtCl$_4$L$_2$	89.4	145.7	1.63

[a]$L = (CH_3)_3{}^{15}N$ or $(CH_3)_3{}^{31}P$. [b]K measured in 10^{20} NA^2m^{-3}

(5.21) Reduced coupling constants in analogous P and N complexes

coupling constants are corrected for differences in nuclear properties, thus facilitating comparisons of coupling to different nuclei). Again the interpretation assumes that the coupling constants measure *s* effects. In the complexes studied, the ratio of reduced coupling constants for P and N complexes is quite different for the Pt(II) and Pt(IV) compounds. If we assume that the bonding in the Pt(IV) case is primarily σ in type, accepting the arguments given above, then the much greater increase in K_{PtP} than in K_{PtN} in going from Pt(IV) to Pt(II) is best interpreted as arriving from a difference in bond type between Pt(II)-N and Pt(II)-P, and this difference is most likely to be the occurrence of significant amounts of π bonding in the phosphorus complex. The enhanced σ bonding seen directly in the coupling constant data in the Pt(II)-P compound is accompanied by increased π interaction in the same compound.

5.5   Reaction mechanisms at phosphorus of C.N.3

Because most C.N.3 phosphorus compounds containing reactive groups attached to phosphorus are labile, there is very little work in this area aside from a few preliminary stereochemical observations. It has been noted, for example, that addition of chloride ion to some RR^{1}PCl compounds is accompanied by coalescence in the nmr of signals arising from previously diastereotopic groups, and this observation is compatible with chloride substitution at phosphorus with inversion of configuration (5.22).

CH$_{3a}$, trans to P-Cl; CH$_{3b}$ cis to P-Cl

(5.22)   Nucleophilic substitution at C.N.3 P with inversion

More convincing demonstrations of inversion accompanying nucleophilic substitution have been the reactions of nucleophiles with a chloro-substituted phosphetan and of organometallic reagents with a phosphine (5.23).

$(+) - (R)_p$

$(+) - (S)_p$
complete inversion

(5.23)   Inversion accompanying nucleophilic substitution at C.N.3P

Although the stereochemistry of these reactions is now established, reaction mechanisms are still obscure. Of particular interest is the question of whether C.N.4 intermediates are involved.

5.6   Redistribution

A commonly observed process in the chemistry of C.N.3 compounds with more electronegative substituents such as halogen, OR, NR_2, on phosphorus is the redistribution reaction, in which there is an exchange of substituents between phosphorus atoms (5.24). There is some information on the thermo-dynamic parameters of these processes, but here, too, little is known mechanistically.

(5.24) $$PCl_3 + P[N(CH_3)_2]_3 \rightleftharpoons PCl_2N(CH_3)_2 + PCl[N(CH_3)_2]_2$$
$$CH_3PBr_2 + CH_3PCl_2 \rightleftharpoons CH_3PClBr$$

Many of these redistribution reactions are extremely rapid and exothermic, and a nonrandom product distribution is achieved, especially when nitrogen substituents are involved (5.25).

$$\tfrac{2}{3}P[N(C_2H_5)_2]_3 + \tfrac{1}{3}PCl_3 \longrightarrow [(C_2H_5)_2N]_2PCl$$
$$\Delta H = -36 \text{ kJ}$$

$$PCl_3 + [(C_2H_5)_2N]_2PCl \rightleftharpoons 2(C_2H_5)_2NPCl_2$$

$$K = \frac{[(C_2H_5)_2NPCl_2]^2}{[PCl_3][((C_2H_5)_2N)_2PCl]} = 2.5 \times 10^7$$

(5.25)   Equilibrium and enthalpy in redistribution

5.7   Oxidation

Many oxidation reactions of C.N.3 phosphorus compounds are known with molecular oxygen, halogens, and other oxidants. These are often rapid (sometimes explosive) reactions, and their driving force can be recognized as the stability of phosphorus in higher coordination states, and the strengths of the new bonds formed. Some oxidation reactions are given in (5.26). The reactions are often of high yield and can be preparatively useful; however, few studies have been made of oxidation mechanisms of phosphorus C.N.3 compounds. In the studies that are available, both free-radical and ionic mechanisms have been identified.

$$PCl_3 \xrightarrow[\text{reflux}]{O_2} POCl_3 \quad \text{(used industrially)}$$

$$(nC_4H_9)_3P \xrightarrow[\text{acetone}]{H_2O_2} (nC_4H_9)_3PO$$

$$(C_2H_5O)_3P \xrightarrow{N_2O_4} (C_2H_5O)_3PO$$

$$(C_6H_5)_3P \xrightarrow{Br_2} (C_6H_5)_3PBr_2$$

$$(C_6H_5O)_3P \xrightarrow{O_3} (C_6H_5O)_3PO$$

$$(C_6H_5)_3P \xrightarrow{S_8} (C_6H_5)_3PS$$

$$R_3P \xrightarrow{R'N_3} R_3PNR'$$

(5.26)   Some preparatively useful oxidations

For example, in the reaction between C.N.3 compounds and dialkyl peroxides, C.N.4 phosphoranyl radicals have been detected by esr (Section 3.2.4) and a reaction mechanism involving these intermediates is proposed (5.27):

$$ROOR \xrightarrow[\text{heat}]{h\nu \text{ or}} 2RO\cdot$$

$$R'_3P + RO\cdot \longrightarrow R_3\dot{P}OR \quad \text{phosphoranyl radical}$$

$$R_3\dot{P}OR \longrightarrow R_3PO + R\cdot$$

$$R\cdot \longrightarrow \text{termination products (e.g., } R_2)$$

(5.27)   Mechanism of dialkyl peroxide oxidation

However, labeling experiments on the similar-appearing reaction with dibenzoyl peroxide strongly support an ionic mechanism for this reaction (5.28):

$$R_3P: \longrightarrow \begin{array}{c} OC(O)Ph \\ | \\ OC(O)Ph \end{array} \longrightarrow \begin{array}{c} R_3\overset{+}{P}O \frown C(O)Ph \\ PhCO_2^- \end{array}$$

$$\longrightarrow R_3PO + PhC(O)OC(O)Ph$$

(5.28)   Mechanism of dibenzoyl peroxide oxidation

The industrially important aereal oxidation of PCl_3 to $POCl_3$ has been characterized as a radical-chain process with chlorine atoms as the principal chain carriers (5.29):

$$Cl \cdot + PCl_3 \longrightarrow \dot{P}Cl_4$$

$$\dot{P}Cl_4 + O_2 \longrightarrow \dot{P}Cl_4O_2$$

$$\dot{P}Cl_4O_2 + PCl_3 \longrightarrow 2POCl_3 + Cl \cdot$$

(5.29)   Radical-chain oxidation of PCl_3

The nature of the initiator in this reaction is still not certain.

6   Phosphorus of C.N.4

6.1   Notes on synthesis

6.1.1   From C.N.3 to C.N.4

As already discussed (Section 5.7), the facile oxidation reactions of C.N.3 compounds often lead in a preparatively useful way to C.N.4 compounds. In addition to these already-cited reactions, there are other generally useful transformations that have the same overall results, and two of these will be discussed now.

Michaelis-Arbuzov reaction. In the Michaelis-Arbuzov reaction, as shown in (6.1), a C.N.3 compound carrying a P–OR group is converted into a C.N.4 compound with a P:O group. A new carbon–phosphorus bond is formed concurrently. The reaction is of wide scope and applicability and is very useful in synthesis (6.2). The mechanism is well understood. A quasiphosphonium intermediate (6.3), which can be isolated in some instances, undergoes nucleophilic attack at carbon by the anion to give the phosphoryl product.

$$G_2P-OR + R'X \longrightarrow G_2P(:O)R' + RX$$

(6.1)   The Michaelis-Arbuzov reaction. If $R = R'$, the process is an isomerization. G can be alkyl, aryl, OR, NR_2, and so on. R' must generally be alkyl (or hydrogen); primary reacts faster than secondary; tertiary alkyl halides do not undergo this reaction.

$$(CH_3O)_3P \xrightarrow[\text{1 mole \% CH}_3I]{\text{reflux}} (CH_3O)_2P(:O)CH_3$$

$$C_6H_5P(OCH_3)_2 + C_2H_5I \longrightarrow C_6H_5P(:O)(OCH_3)C_2H_5 +$$

$$C_6H_5P(:O)(OCH_3)(CH_3) + CH_3I$$
a secondary product from the CH_3I produced

$$(C_6H_5)_2POCH_3 + CF_2{:}CCl\ CClF_2 \longrightarrow (C_6H_5)_2P(:O)CF_2CCl{:}CF_2 + CH_3Cl$$

(cont.)

$$(C_2H_5O)_3P + HCl \longrightarrow (C_2H_5O)_2P(:O)H + C_2H_5Cl$$

(6.2)   Examples of the Michaelis-Arbuzov reaction

$$G_2P: \longrightarrow R'\!\!-\!\!X \longrightarrow G_2\overset{+}{P}R'$$

$$\underset{RO}{|} \qquad\qquad\qquad \underset{R\!-\!O}{|}$$

$$X^-$$

Quasi-phosphonium
intermediate; isolable
when, for example, G = PhO,
R = Ph, R'X = CH₃I

$$\longrightarrow G_2PR' + RX$$
$$\underset{O}{\overset{\|}{}}$$

(6.3)   Mechanism of the Michaelis-Arbuzov reaction

An important variant of the reaction, known as the Perkow reaction, employs halogen-substituted ketones and leads to enol-phosphate end products, such as the insecticide phosdrin (6.4).

$$(CH_3O)_3P + CH_3COCHClCO_2C_2H_5 \longrightarrow (CH_3O)_2P(:O)C(CH_3):CHCO_2C_2H_5$$

phosdrin (cis/trans mixture)

(6.4)   Preparation of the insecticide phosdrin

Kinnear-Perren reaction. Another method of forming a new carbon–phosphorus bond is afforded by the Kinnear-Perren reaction, which involves an alkyl halide, phosphorus trichloride [or a substituted phosphorus(III) chloride], and an aluminum chloride catalyst. Examples of the process are given in (6.5).

$$RCl + PCl_3 + AlCl_3 \longrightarrow R\overset{+}{P}Cl_3\; AlCl_4^-$$
$$\xrightarrow{H_2O} RP(:O)Cl_2$$

Examples:

$$C_2H_5Cl \xrightarrow[\text{(b) H}_2\text{O}]{\text{(a) PCl}_3\text{; AlCl}_3} C_2H_5P(:O)Cl_2$$

$$(CH_3)_2CHCl \xrightarrow[\text{(b) H}_2\text{O}]{\text{(a) PhPCl}_2\text{, AlCl}_3} (Ph)\,[(CH_3)_2CH)]\,P(:O)Cl$$

(6.5)   Kinnear-Perren reaction

6.1.2   C.N.4 interconversions

Interconversions of C.N.4 compounds provide routes to many different types of new C.N.4 compounds. The scope of such processes is very wide; halogen attached to phosphorus of C.N.4 is a very versatile leaving group and can be alkylated or arylated by organometallic reagents, or displaced by many other nucleophiles. Some typical reactions are illustrated in (6.6).

$$C_2H_5P(:O)Cl_2 + 2C_6H_5S^- \longrightarrow C_2H_5P(:O)(SC_6H_5)_2$$

$$\xrightarrow{2(CH_3)_2NH} C_2H_5P(:O)(Cl)[N(CH_3)_2] +$$

$$(CH_3)_2NH_2^+Cl^-$$

$$\xrightarrow{2nC_4H_9Li} C_2H_5P(:O)(nC_4H_9)_2$$

$$\xrightarrow{2CH_3O^-} C_2H_5P(:O)(OCH_3)_2$$

$$(C_2H_5O)_2P(:O)H + Cl_2 \longrightarrow (C_2H_5O)_2P(:O)Cl + HCl$$

$$\downarrow KF, benzene$$

$$(C_2H_5O)_2P(:O)F + KCl$$

$$CH_3P(:O)(OCH_3)_2 + PCl_5 \longrightarrow CH_3P(:O)Cl_2$$

(6.6)   C.N.4 Interconversions

6.2   Stereochemical aspects

The tetrahedral geometry of C.N.4 phosphorus was indicated, long before physical methods confirmed it in hosts of compounds, by the resolution of a chiral phosphine oxide, $(CH_3)(C_2H_5)(C_6H_5)P{:}O$. Phosphine oxides are Lewis bases, coordinating through oxygen, and are basic enough to form stable salts with strong acids. The resolution was carried out using the diastereomeric salts formed with (+)bromocamphorsulfonic acid. Other phosphine oxides and sulfides were resolved but curiously, and for reasons that are still not clear, resolution of chiral phosphonium salts was not achieved until much later. This resolution was a key step in the preparation of chiral phosphines, as has already been mentioned (Section 5.2). Resolved optically active phosphonium salts are generally stable to racemization.

Absolute configurations of a number of chiral C.N.4 compounds have been determined by X-ray diffraction methods. For example, the enantiomer of the $(CH_3)(nC_3H_7)(C_6H_5)(C_6H_5CH_2)P^+$ cation that has a positive rotation at the sodium D line is configurationally S (6.7).

(6.7)   Absolute configuration (S) of a chiral phosphonium salt

Configurational correlations have been made in phosphorus chemistry by methods quite analogous to those established in carbon chemistry, and a number of these will be discussed in the next section when particular reactions are examined.

6.3 Reactions and reaction mechanisms

6.3.1 Reactions of phosphonium salts with nucleophiles

Phosphonium salts are much more diverse in their reaction patterns with nucleophiles than are ammonium salts. This stems, in part, from the ability of phosphorus to accommodate C.N.5 intermediates. In this section we shall examine four different types of these reactions, among them the formation of phosphorus ylids, perhaps the most widely used class of phosphorus compounds in organic synthesis.

Nucleophilic substitution at α-carbon [$S_N(C)$]. In nucleophilic substitution at the α-carbon the reaction process is

(6.8) $R_3\overset{+}{P}R'X^- \longrightarrow R_3P + R'X$

which is the reverse of the quaternization reaction. This is a relatively unusual reaction pathway and is favored by R' groups that do not stabilize other possible intermediates (see following sections) and by X^- ions of low basicity or nucleophilicity, such as Cl^-. The reaction is performed pyrolytically.

E_2 elimination. Treatment of compounds of the type $R_3\overset{+}{P}CH_2CH_2G$ with bases may lead to E_2 elimination if the group G is capable of stabilizing a developing negative charge at the carbon atom α to it. Two variants of the general elimination reaction are shown in (6.9) together with plausible mecha-

$$R_3\overset{+}{P}-CH_2CHCN \longrightarrow R_3P \quad CH_2\!\!=\!\!CHCN$$
$$HO \quad H \qquad\qquad\qquad H_2O$$

Base elimination from a β-cyanoethylphosphonium salt

$$R_3\overset{+}{P}-CH_2O-H \quad \overset{-}{O}H \longrightarrow R_3P \quad CH_2O \quad H_2O$$

Base elimination from a hydroxymethylphosphonium salt

(6.9)   Elimination reactions

nisms. Both reactions are useful in that they provide a reversal of an alkylation process, and so can serve in the directed synthesis of various tertiary phosphines. An example of the use of the β-cyanoethyl group in this manner is shown in (6.10).

$$C_6H_5PH_2 \xrightarrow[\text{base}]{2CH_2:CHCN} C_6H_5P(CH_2CH_2CN)_2$$

$$\xrightarrow{CH_3I} C_6H_5\overset{+}{P}(CH_3)(CH_2CH_2CN)_2I^-$$

$$\xrightarrow{HO^-} CH_2:CHCN + C_6H_5(CH_3)P(CH_2CH_2CN)$$

$$\xrightarrow{nC_3H_7Br} C_6H_5(CH_3)\overset{+}{P}(nC_3H_7)Br^-$$
$$\underset{\begin{array}{c}\text{(resolvable via}\\\text{diastereomeric salts)}\end{array}}{\overset{|}{CH_2CH_2CN}}$$

$$\xrightarrow{OH^-} \underset{\text{optically active}}{C_6H_5(CH_3)(nC_3H_7)P} + CH_2:CHCN$$

(6.10)   Use of CH_2CH_2CN group in phosphine synthesis

Nucleophilic displacement at phosphorus ($S_N(P)$). Nucleophilic displacement at phosphorus,

(6.11) $R_3\overset{+}{P}R' + OH^- \longrightarrow R_3P{:}O + R'H$

is the most common type of base reaction. In terms of leaving-group ability, the order has been established as $C_6H_5CH_2 > C_6H_5 > CH_3$. The detailed mechanism of this reaction has been established by a variety of studies. The reaction is first order in phosphonium salt and second order in hydroxide ion (and is thus third order overall). In the series of reactions (6.12), the partial rate constants for the loss of the Y-substituted benzene ring give an excellent Hammett $\sigma{-}\rho$ plot with $\rho = 3.64$, showing the large electron demand at phosphorus in the process. The stereochemistry has been established for some systems as 100% inversion at phosphorus by means of the cycle outlined in (6.13) (although in other systems the stereochemical results are less clean). Putting all the evidence together, the plausible mechanism in (6.14) emerges, in which

(6.12)

a C.N.5 intermediate is involved. This may eliminate directly with inversion, as shown, or in some cases may undergo ligand permutation (Section 7.3). As we shall see in Section 7.3, there are many reactions of C.N.4 phosphorus compounds in which C.N.5 intermediates are implicated.

(6.13)   Stereochemistry of OH^- attack at P^+

Kinetic pattern:

$$R_3\overset{+}{P}R' + OH^- \underset{\text{equilibrium}}{\overset{\text{fast}}{\rightleftharpoons}} R_3P(R')(OH) \underset{\text{fast equilibrium}}{\overset{OH^-}{\rightleftharpoons}} R_3P(R')O^-$$

$$\xrightarrow[\text{rate limiting}]{\text{slow}} R_3P:O + R'^- \xrightarrow[\text{fast}]{\text{solvent}} R'H$$

Stereochemistry:

(6.14)   Mechanism of OH^- attack at P^+

Alkylene phosphorane (ylid) formation. Phosphonium salts that have α-hydrogen atoms may react with strong bases via an α-elimination route to give alkylene phosphoranes (phosphonium ylids) (6.15):

(6.15) $R_3\overset{+}{P}CHR'R'' \xrightarrow{B} R_3PCR'R'' + BH^+$

The bonding situation in these compounds has previously been examined in some detail (Section 4.3) and, in the present author's judgment, unless potentially conjugating groups are present, the compounds are best regarded as alkylene phosphoranes, with the recognition that the P:C double bond in these compounds will be quite polar in the sense $\overset{\delta+\delta^-}{P:C}$. The reader should be aware, however, that the alternative description of these compounds as phosphonium ylids is the dominant one in the chemical literature.

The ease of formation of the alkylene phosphorane depends on the acidity of the α-hydrogen to be removed. In acyl phosphonium salts, very weak bases may suffice; the acylated phosphoranes produced are relatively robust, insensitive compounds. In alkylphosphonium salts, strong bases are needed and the phosphoranes thus produced are sensitive and reactive compounds (6.16). Some alternative methods of preparing alkylene phosphoranes, though unrelated to base elimination reactions, are included here for completeness, because some of the chemistry of these compounds will be discussed below (6.17).

$(C_6H_5)_3P + BrCH_2COC_6H_5 \longrightarrow (C_6H_5)_3\overset{+}{P}CH_2COC_6H_5Br^-$

$\xrightarrow[H_2O]{Na_2CO_3} (C_6H_5)_3P:CHCOC_6H_5$

$(C_6H_5)_3\overset{+}{P}CH_3I^- \xrightarrow{nC_4H_9Li} (C_6H_5)_3P:CH_2 + LiI + C_4H_{10}$

(6.16)   Preparation of alkylene phosphoranes

Staudinger reactions:

$R_3P + RR'CN_2 \longrightarrow [R_3PN_2CRR']$

$\longrightarrow N_2 + R_3P:CRR'$

Carbenoid precursors:

$R_3P + CH_2Cl_2 \xrightarrow{tBuOK} [R_3P:CHCl]$
                                    Usually not
                                    isolated, but
                                    allowed to react
                                    further in situ

"Ylid" alkylation or acylation:

$$R_3P:CHCO_2CH_3 \xrightarrow{C_6H_5CH_2Br} R_3P:C(CO_2CH_3)(CH_2C_6H_5)$$

$$C_6H_5COSEt \searrow$$

$$R_3P:C(CO_2CH_3)(COC_6H_5)$$

(6.17)   Some preparations of alkylene phosphoranes

Reactions of alkylene phosphoranes (phosphonium ylids). Alkylene phosphoranes are very reactive compounds, and are readily hydrolyzed to yield phosphine oxides (6.18):

(6.18) $R_3P:CR'R'' + H_2O \longrightarrow R_3P:O + H_2CRR'$

They can be alkylated or acylated as shown in (6.17). Their most important reaction, however, is that with aldehydes or ketones to yield alkenes, a reaction discovered by Staudinger and rediscovered by Wittig, and generally known as the Wittig reaction. A general formulation of the reaction and two examples of it are given in (6.19). The Wittig reaction has been exceedingly

$$R_3P:CR'R'' + R'''COR''' \longrightarrow R_3P:O + R'R''C:CR'''R'''$$

(6.19)   The Wittig reaction in action

successful in the synthesis of natural products and strained alkenes, and there is now a large literature devoted to it, which can be accessed through the Bibliography.

Mechanisms and intermediates in Wittig reactions. Because of its synthetic importance, the Wittig reaction has been carefully studied. Most of the physicochemical work, however, has been with atypical alkylene phosphoranes, that is, compounds that are relatively robust and that react slowly, allowing kinetic study and even the isolation of intermediates. A general scheme is shown in (6.20), but there is no general agreement as to whether both betaines and 1,2-

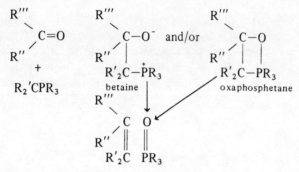

(6.20) General mechanism for Wittig reactions

oxaphosphetanes are necessarily involved in all Wittig reactions. In some reactions isolable betaines are formed (6.21), in others oxaphosphetanes are impli-

$$(C_6H_5)_3P:C(CH_3)_2 + (C_6H_5)_2C:C:O \longrightarrow (C_6H_5)_3\overset{+}{P}-\underset{|}{C}(CH_3)_2$$

$$O^- - C:C(C_6H_5)_2$$

betaine: $\mu = 4.3D$ $\delta 31_P = +36$ ppm

$$\overset{heat}{\longrightarrow} (C_6H_5)_3P:O + (CH_3)_2C:C:C(C_6H_5)_2$$

(6.21) A betaine intermediate in a Wittig reaction

cated (6.22), but, as was said above, these are in atypical reactions. Similarly, kinetic studies that support the mechanism in (6.20) have been carried out with robust alkylene phosphoranes – for example, $(C_6H_5)_3P:CHCO_2CH_3$ – and indicate a highly negative entropy of activation, indicative of a constrained transition state. The rates of these reactions increase with solvent polarity, which is consistent with a betaine intermediate or betainelike transition state, of higher polarity than the starting alkylene phosphorane.

The stereochemistry of the reaction at phosphorus has been established as

$$(C_6H_5)_3P:C:P(C_6H_5)_3 + (CF_3)_2C:O \longrightarrow \begin{array}{c} (C_6H_5)_3P{-}C:P(C_6H_5)_3 \\ \mid \quad\; \mid \\ O{-}C(CF_3)_2 \end{array}$$

oxaphosphetane

$$\delta\,31_P = -7.33 \text{ ppm and } +54.0 \text{ ppm (high field:ring P)}$$
$$^2J_{PP'} = 47 \text{ Hz}$$

$$\xrightarrow{\text{heat}} (C_6H_5)_3P:O + (C_6H_5)_3P:C:C(CF_3)_2$$

(6.22)   An oxaphosphetane intermediate in a Wittig reaction

100% retention, by the use of chiral phosphines (6.23), which agrees with the mechanism outlined in (6.20) in which there is cis-fragmentation of either a betaine or an oxaphosphetane.

(6.23)

6.3.2   Attack of nucleophiles at C.N.4 phosphorus carrying a leaving group

It is perhaps in this area that the complex potentialities of reaction mechanisms at phosphorus centers can be best and most fully demonstrated. A variety of mechanisms is possible, even restricting the discussion, as we shall in this section, to ionic reactions. And there is evidence for many different mechanisms depending on the system involved. Because the intimate details of some mechanisms involve a consideration of C.N.5 compounds, in this section we shall discuss only two types of reaction: those involving an elimination-addition sequence, and those apparently involving direct displacement. This division of an important class of reactions is not altogether satisfactory, and involves some artificialities, but it does have the advantage of rationally subdividing the topic.

Elimination-addition sequence. The formal representation of the elimination-addition process is shown in (6.24) and involves an intermediate that is either a metaphosphate (E = Y = O; G = RO) or a metaphosphate analog.

metaphosphate
analog

(6.24) The elimination-addition mechanism

It is worth pointing out that stable C.N.3 metaphosphate analogs are now known (2.9). There is also excellent evidence, primarily from trapping experiments, that monomeric methyl metaphosphate (intermediate in G = CH_3O, E = Y = O) can be prepared in the gas phase as a highly reactive species (6.25).

(6.25) Production of methyl metaphosphate

There is much indirect kinetic and product evidence for the participation of metaphosphate analogs in nucleophilic reactions. Thus, in the nucleophilic attacks on phosphorus amides presented in (6.26), the great difference in rates for hydroxide, which is a poor nucleophile but a good base in aqueous solu-

$$(nC_3H_7NH)_2P(:O)Cl \xrightarrow[K_1]{H_2O,\ B} (nC_3H_7NH)_2P(:O)OH$$

$$[(CH_3)_2N]_2P(:O)Cl \xrightarrow[K_2]{H_2O,\ B} [(CH_3)_2N]_2P(:O)OH$$

For B = pyridine, F^-, $K_1 \approx K_2$

For B = OH^-, $K_1 \approx 10^6 K_2$

(6.26)   Elimination–addition reactions

tion, is strongly suggestive of a conjugate base mechanism like the one shown. This mechanism is available only when there is a proton on nitrogen, and leads to fast reaction because the intermediate conjugate base can very readily lose chloride ion.

Another kinetic demonstration of this type of mechanism comes from a study of the hydrolysis of the phosphagens [phosphorylated guanidines (6.27)]. Kinetically the hydrolysis reaction is first order in phosphagen. The

(6.27)   Phosphagen hydrolysis

pH dependence of the rate constant is bell shaped, with a flat maximum at pH 1.5–2.5. The pH dependence curve can be fitted exactly by the scheme

(6.28) $AH_2^+ \xrightleftharpoons{K_{a_1}} H^+ + AH \xrightleftharpoons{K_{a_2}} H^+ + A^-$

in which *only* the AH species is hydrolyzed. The kinetically derived values of pK_{a_1} and pK_{a_2} are equal to those determined by direct potentiometric titration. The entropy of activation is close to zero, which is indicative of a unimolecular process (6.29). Because the neutral zwitterion is the only hydrolytically active species, this is consistent with an elimination–addition mechanism with intramolecular proton transfer [(6.29) with $R' = H$]. The importance of the proton transfer is indicated by the total lack of reactivity of the benzyl ester [(6.29) with $R' = C_6H_5CH_2$].

(6.29) Mechanism of phosphagen hydrolysis

These phosphagen hydrolyses are of more than simple mechanistic and kinetic importance. The in vivo phosphorylation of ADP during periods of muscular effort is by phosphocreatine (6.30), for which the phosphagens are good models; and hydrolysis is the phosphorylation of H_2O. So the phosphagen reactions may be informative models for a very important biochemical process.

(6.30) Phosphorylation of ADP by phosphocreatine (stoichiometry only)

Direct displacement. It must be said at once that there is no firm evidence for a direct displacement mechanism operating in nucleophilic attack at phosphorus C.N.4 compounds. The problem is as follows. Kinetic information, such as the existence of a second-order reaction, tells us only the composition of the transition state *or intermediate,* but cannot distinguish between direct displacement and a process involving, as rate-limiting step, the formation of a genuine C.N.5 intermediate (6.31).

Occam's razor cuts both ways in this instance. It could be argued that where there is no evidence for an intermediate, it should not be invoked. However, because the intervention of a C.N.5 intermediate is strongly implied in many

$$\geq P-N+X$$

$$\geq P-X+N \quad \xrightarrow[\text{order}]{\text{second}} \quad K_2 \quad K_{-2}$$

$$K_1 \quad K_{-1}$$

$$\geq P \underset{N}{\overset{X}{<}}$$

(6.31) Alternative explanations for a second-order process

reactions of this class (see Section 7.3), it could also be argued that the same overall kinetic scheme should be applied to all such displacements, and that an intermediate should be inferred even when it cannot be demonstrated. This author's personal judgment is to reject the latter argument and accept the former, simpler, view on the evidence that is presently available. There are quite a number of nucleophilic reactions involving second-order kinetics that, stereochemically, show 100% inversion at phosphorus, and that can be considered as direct displacement reactions.

There are, for instance, elegant double-labeling experiments which indicate that in these particular substitutions every act of substitution gives rise to inversion at phosphorus. There are also a number of Walden inversion cycles that indicate, at least, predominant inversion at phosphorus along with second-order kinetics – for example, (6.32).

$$\underset{\underset{C_2H_5}{\overset{O}{\underset{|}{\overset{\|}{P}}}}}{\overset{}{\underset{OCH_3}{}}} \quad + {}^{14}CH_3O^- \rightleftharpoons (C_6H_5)(C_2H_5)({}^*CH_3O)P:O$$

C_6H_5

racemic, labeled

↑

resolved, optically
active

↓

K (label incorporation for both reactions) =
0.5K (polarimetric)

$$\underset{\underset{C_2H_5}{\overset{S}{\underset{|}{\overset{\|}{P}}}}}{\overset{}{\underset{OC_2H_5}{}}} \quad + {}^{36}Cl^- \rightleftharpoons (C_2H_5)(C_2H_5O)({}^*Cl)P:S$$

inversion at every act of substitution at P   94

(6.32) A Walden inversion cycle at P

There are also some very informative reactions on cyclic systems containing phosphorus (6.33). In these reactions, mixed stereochemical results are observed and can be summarized as follows:

X = O, S

(6.33) Nucleophilic attack on a cyclic system

In this series, when N is strongly basic (e.g., $^-$SR, $^-$OR) and Y is a good leaving group (e.g., Cl), inversion at phosphorus predominates and the reaction may be a direct displacement. When N is less basic (e.g., ROH) and Y is a poor leaving group (e.g., R$_2$N), then retention at phosphorus predominates, and C.N.5 intermediates may be involved. These will be discussed in section 7.3.

6.4 Catenated C.N.4 compounds

Much of the inorganic chemistry of C.N.4 phosphorus is concerned with catenated compounds of various types. In this section just a few of the more important types will be touched upon to indicate the scope of the subject. Detailed reviews of many of these topics are available (see the Bibliography).

The cyclophosphazenes, $(G_2PN)_n$, have already been discussed in some detail (Section 4.3.5), and all that remains to be added at this point is that there is also an extensive family of linear phosphazenes. For instance, in the reaction between NH_4Cl and PCl_5, in addition to cyclophosphazenes $(Cl_2PN)_n$, a range of compounds of empirical formula $(Cl_2PN)_n \cdot PCl_5$ is produced, which have been shown to be the linear compounds $Cl_3P{:}N(Cl_2P{:}N)_n PCl_3^+ Cl^-$. When n is large, these polymers are elastomers (although they are subject to hydrolytic and thermal degradation), and there has been much research, with some success, into the preparation of stable phosphazene derivatives that might serve as high-temperature elastomers.

The phosphorus–oxygen system also leads to many catenated compounds, both linear and cyclic. (For simplicity, only the fully ionized formulas will be drawn in this section.) Thermal condensation by dehydration of various hydrogen phosphates, HPO_4^{2-}, dihydrogen phosphates, $H_2PO_4^-$, and mixtures of these compounds gives rise to a range of linear polyphosphates $[O_3P{\text{-}}(OPO_2)_n PO_3]^{(n+4)-}$, from which individual species can sometimes be isolated and characterized. Thus diphosphates, $(O_3P{-}O{-}PO_3)^{4-}$ (which used to be known as pyrophosphates), triphosphates, $[O_3P{\text{-}}(OPO_2){\text{-}}OPO_3]^{5-}$ and tetraphosphates are well known and characterized. Linear polymers with molecular weights in excess of 10^6 are known in this system.

The condensation reactions also give rise to cyclopolyphosphates, $\begin{smallmatrix} O \\ \| \\ {\text{-}}(OP){\text{-}}_n \\ | \\ O^- \end{smallmatrix}$

(formerly known as metaphosphates), in which individual ring compounds from $n = 3$ to $n = 8$ have been isolated and fully characterized. Chromatographic methods suggest the presence of larger rings as well.

Reactions of polyphosphates with tetraphosphorus decaoxide, P_4O_{10}, lead to the class of ultraphosphates, which can be regarded as cross-linked polyphosphates (6.34):

(6.34) Ultraphosphate structural segment

The ultimate structure in this cross-linked area is that of P_4O_{10} itself (6.35):

$$O$$
$$\|$$
$$P$$

(6.35)   Structure of P_4O_{10} (Td symmetry)

Because $-NH-$ is isolectronic with $-O-$, it is not surprising that imido-phosphates of many kinds also exist. And when we ring the many changes possible substituting $-S-$ for $-O-$, $=S$ for $=O$, or, partially, As or Sb for P, we can see that an enormously large area of chemistry is encompassed in this topic of catenated compounds.

6.5   Reduction

As one would expect from a consideration of bond strengths, the reduction of C.N.4 compounds containing P–O bonds to C.N.3 compounds can only be ef-fected by powerful reducing agents. However, this is a very desirable reaction. For instance, in the synthesis of compounds containing carbon–phosphorus bonds, the most powerful and general methods (Michaelis-Arbuzov or Kinnear-Perren reactions) give C.N.4 products, but often C.N.3 compounds related to these are desired for study or for use as donors. Then, in the field of chiral compounds, resolution of active compounds is usually easier for stable, ro-bust C.N.4 compounds than for the often easily oxidized C.N.3 compounds.

Consequently, much research has gone into the choice of reducing agents for this process. Among the most powerful agents is lithium tetrahydroalu-minate, which can carry out many reductions such as those shown in (6.36).

$$CH_3P(:O)(OCH_3)_2 \xrightarrow{\text{LiAlH}_4} CH_3PH_2$$
$$C_6H_5P(O)Cl_2 \xrightarrow{\text{LiAlH}_4} C_6H_5PH_2$$
$$(C_6H_5)_2P(O)Cl \xrightarrow{\text{LiAlH}_4} (C_6H_5)_2PH$$
$$(CH_3)_2P(O)C_6H_5 \xrightarrow{\text{LiAlH}_4} (CH_3)_2(C_6H_5)P$$

(6.36)   Reductions with lithium tetrahydroaluminate.

Although many such reductions are reported, there has really not been a sys-tematic study of this class of reactions. Yields are often low, and it would seem that an examination of modified hydroaluminates, or of some of the recently prepared B–H reducing agents, in the area of phosphorus chemistry would be very useful.

In the limited field of reduction of chiral phosphine oxides to optically active phosphines, outstanding successes have been achieved with reductants containing silicon. Although hexachlorodisilane, Si_2Cl_6, has been widely used, it appears that the reagent of choice at the present time is phenylsilane, $C_6H_5SiH_3$, which effects reduction with 100% retention of configuration at phosphorus. This reaction is probably a four-centered process with a cyclic transition state (6.37).

(6.37)   Reduction of P:O by phenylsilane

7 Phosphorus of C.N.5

7.1 Stereochemical aspects; synthesis

In the initial preliminary survey of C.N.5 (Section 2.5), it was pointed out that most compounds in this category have a trigonal bipyramidal (t.b.p.) geometry, although some examples of square pyramidal (s.p.) geometry have recently been described (2.11, 7.1). With suitable substituents, either geometry could,

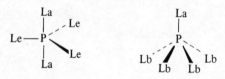

t.b.p. (trigonal bipyramid) s.p. (square pyramid)
La = axial ligand   La = apical ligand
Le = equatorial ligand   Lb = basal ligand

(7.1)   Ideal geometries of C.N.5

in principle, give rise to chiral resolvable compounds but in fact, as we shall see, there is a relatively small energy difference between these geometries, and interconversions of C.N.5 geometries are facile. So far there are no reported resolutions of chiral C.N.5 compounds. The synthesis of C.N.5 compounds is often based on oxidative addition to C.N.3 compounds. Examples include the halogenation and peroxidation reactions shown in (7.2). Another type of oxi-

$$PF_3 + F_2 \longrightarrow PF_5$$

$$PCl_3 + Cl_2 \longrightarrow PCl_5 \quad \text{(C.N.5 under some conditions)}$$

$$(CF_3)_3P + Cl_2 \longrightarrow (CF_3)_3PCl_2$$

$$(C_2H_5O)_3P + C_2H_5OOC_2H_5 \longrightarrow (C_2H_5O)_5P$$
$$\text{(very unstable)}$$

(7.2)   Oxidative addition yielding C.N.5 compounds

dative addition, shown in (7.3), is that of dicarbonyl compounds to C.N.3 compounds leading to oxophosphoranes. Interconversions, as illustrated in (7.4), are also useful in the preparation of C.N.5 compounds.

$$(CH_3O)_3P + CH_3COCOCH_3 \longrightarrow (CH_3O)_3P \underset{O}{\overset{O}{<}} \overset{CH_3}{\underset{CH_3}{>}}$$

(7.3)   Oxidative addition of a dicarbonyl compound

$$PCl_5 \xrightarrow{\text{SbF}_3} PCl_4F, PCl_3F_2, PCl_2F_3, PClF_4, PF_5$$

$$\xrightarrow[\text{CH}_3\diagdown_{\text{N}}\diagup^{\text{CH}_3}]{\text{C}_6\text{H}_5\text{OH}} (C_6H_5O)_5P$$

$$PF_5 \xrightarrow{\text{NH}_3} H_2NPF_4$$

(7.4)   Interconversions of C.N.5 compounds

7.2   Ligand mobility

Different physicochemical techniques for structure determination have very different time scales for interaction with molecules. This is a consequence of the different energies carried by a "photon" in different regions of the electromagnetic spectrum, and of the application of the uncertainty principle. In general, the higher the photon energy, the shorter the time scale for interaction. We can liken this phenomenon to the ability of a photographer to choose a range of shutter speeds in taking a picture. The longer the shutter is open, the more chance there is of a blurred snapshot resulting from movement of the subject.

The "shutter speeds" of some familiar structural techniques are roughly as follows: X-ray or electron diffraction, of the order of 10^{-20} s; infrared spectroscopy, of the order of 10^{-13} s; nmr spectroscopy, of the order of 10^{-8} to 10^{-1} s (depending on the nucleus observed and the actual system studied).

Let us bring this abstract discussion into focus by considering the application of these techniques to the PF_5 molecule. Electron diffraction, with its very fast speed (10^{-20} s) reveals a t.b.p. structure with longer axial (158 pm) than equatorial (154 pm) P–F bond lengths. The infrared spectrum, with rather slower speed (10^{-13} s), confirms a molecule of D_{3h} symmetry, as shown by electron diffraction. But in the ^{19}F nmr spectrum down to the lowest accessible temperature, only one ^{19}F chemical shift is seen, which implies (barring accidental coincidence of the chemical shifts for axial and equatorial fluorine nuclei which, as we shall see below, is unlikely) that axial and equatorial fluorine nuclei are changing places in this molecule. The observation of a $^1J_{PF}$ in PF$_5$ shows that this process must be occurring intramolecularly.

Phosphorus pentafluoride was, historically, the first example of ligand mobility to be detected in a C.N.5 phosphorus compound, but the phenomenon is a very general one, and ligand mobility is the rule, rather than the exception, for such compounds. For example, in the compound PH$_2$F$_3$, only one ^{1}H chemical shift and one ^{19}F chemical shift are seen at room temperature, whereas below $-60°$C two distinct ^{19}F chemical shifts in the area ratio 2:1 develop. The coupling pattern is compatible with the low-temperature structure that has one equatorial and two axial fluorines (7.5). At $-46°$C the two

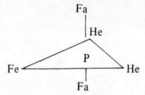

(7.5)   Low-temperature structure of PH$_2$F$_3$

fluorine signals coalesce, and above that temperature only one ^{19}F averaged chemical shift is seen; the PF coupling is retained throughout. A barrier to ligand reorganization of about 38 kJ mol^{-1} can be calculated for this system, and many other such barriers have been determined.

An order of preference for the apical position in the t.b.p. has been established empirically in a range of C.N.5 compounds (this preference is sometimes termed apicophilicity). It is roughly an order of increasing electronegativity, and this can be rationalized by a hybridization argument. In the sp^3d hybridization characteristic of t.b.p., the equatorial bonds are formed with sp^2 hybrid orbitals on phosphorus, and thus have more s character than the axial bonds, which are formed from pd hybrid orbitals. Because s character predominates in bonds to electropositive substituents, it may be expected that these will be formed equatorially, and that more electronegative substituents will prefer axial locations.

The mechanism of ligand interchange in C.N.5 structures has been exhaustively discussed. By a mechanism in this context, as we are discussing an intramolecular phenomenon, is meant the detailed dynamics and the intramolecular pathway by which the interchange occurs. There are two plausible mechanistic suggestions.

The first, known as the Berry mechanism (or sometimes by the very misleading name of "pseudorotation"), involves an s.p. intermediate in scrambling ligands and is best visualized as involving bending modes of the t.b.p. molecule. This mechanism is shown in (7.6).

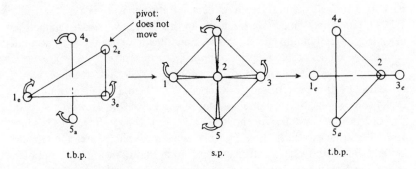

(7.6) Berry mechanism for ligand interchange

A more recent competitor, known picturesquely as the "turnstile" mechanism, is shown in (7.7) and can be visualized as the turning of one group of three ligands against the other two.

Simple molecular orbital calculations of barriers to ligand interchange give similar results (20–40 kJ mol^{-1}) for either mechanism – results that are close to those observed experimentally. It is fair to say that at the present time there is no overwhelming weight of evidence favoring either mechanism, and that the ingenuity of chemists will be taxed to its limit to devise systems and techniques that can profitably be used to make such a fine distinction. It may well be that both mechanisms may be operative, each in appropriate systems.

p = ligand pair
t = turnstile trio
t_e and t_e' bend, reducing
    angle to 90°
p_a and p_e bend by about 9°

p_e and p_a interchange,
while the turnstile trio
turns in the opposite sense

(7.7) Turnstile mechanism for ligand interchange

7.3   C.N.5 as intermediate in reactions of C.N.4 compounds

The properties, including ligand mobility, of isolable C.N.5 compounds have been used as analogies in trying to understand the behavior of hypothetical C.N.5 intermediates in certain reactions. In this section we shall examine the evidence for such intermediates, and see how they have been used to account for otherwise unexpected features of phosphorus reactions.

The most desirable evidence for the existence of C.N.5 intermediates would, of course, be direct spectroscopic observation of them during reactions of C.N.4 compounds. No such direct evidence is yet available. However, a great deal of less direct evidence for the implication of C.N.5 intermediates has been assembled from careful study of hydrolyses of cyclic phosphates and related systems. For example, it is a striking observation that in the dioxaphospholane shown in (7.8), the rate constant for acid-catalyzed hydrolysis with P—O bond cleavage (which is the main reaction, as confirmed by running the reaction in ^{18}O-labeled water) is 10^8 times that for the acyclic analog, dimethyl phosphate.

Reaction

Relative rate constant, K

$$
\begin{array}{c}
CH_2 \\
| \quad\quad\quad\; O \\
CH_2
\end{array}
\underset{O}{\overset{O}{\diagdown}} P(:O)OH + H_2O \quad \xrightarrow{\;H^+\;} \quad HOCH_2CH_2OP(:O)(OH)_2 \qquad 10^8
$$

$$
\begin{array}{c}
CH_3O \\
\quad\quad\;\diagdown \\
CH_3O \diagup
\end{array}
P(:O)OH + H_2O \quad \xrightarrow{\;H^+\;} \quad CH_3OH + CH_3OP(:O)(OH)_2 \qquad 1
$$

(7.8)   Accelerated hydrolysis in a cyclic system

Perhaps an even more striking observation is that the same cyclic phosphate shows rapid incorporation of ^{18}O label (at a rate 0.2 that of the rate of hydrolysis) into the exocyclic oxygen atoms, whereas the acyclic dimethylphosphate shows no detectable incorporation. Although one might be tempted to argue in terms of a "ring-strain" effect in the accelerated hydrolysis shown in (7.8), it seems harder to argue ring strain in a reaction like the ^{18}O exchange (7.9), in which the ring is not opened.

The ingenious explanation of these effects, due largely to Westheimer, invokes a C.N.5 intermediate that is formed in an accelerated process from the ring-strained cyclic phosphate. The intermediate can then undergo ligand permutation and ligand loss to yield products (7.10).

If, in the intermediate A in (7.10), the axial bond opposite the entering H_2O ligand breaks at a rate greater than that of ligand permutation, then ring opening results (7.11).

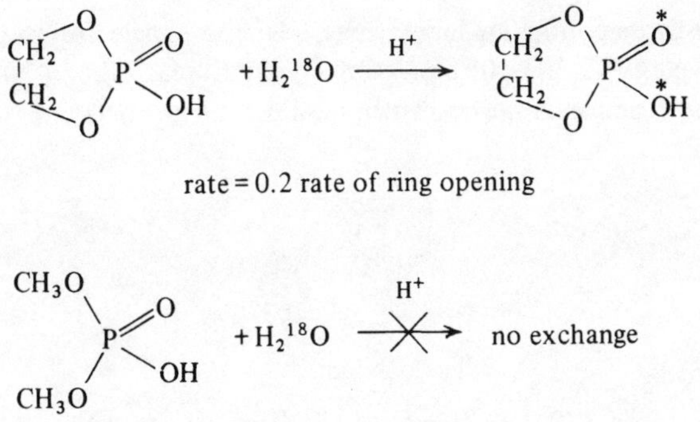

H+
+ H2 18O
rate = 0.2 rate of ring opening
no exchange

(7.9) Accelerated ¹⁸O incorporation into a cyclic system

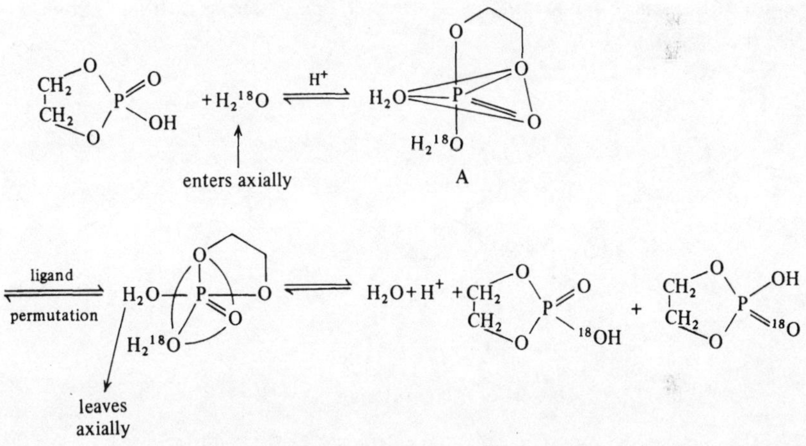

H+
enters axially
A
ligand
permutation
leaves
axially

(7.10) C.N.5 intermediate in ¹⁸O incorporation

A
H+ + ⁻OCH2CH2OP(:O)(OH2)(OH)
HOCH2CH2OP(:O)(OH)2

(7.11) Ring opening of the C.N.5 intermediate

The driving force for the accelerated formation of the intermediate A (7.10, 7.11) in the hydrolysis is the relief of ring strain in the cyclic phosphate (shown to be present both by thermochemical measurements and geometrically) in going to the preferred 90° angle in the C.N.5 intermediate. In acy-

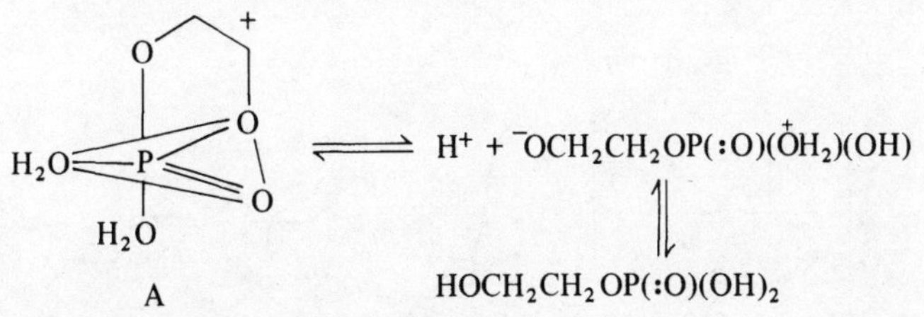

clic systems or those containing larger rings, where the angle at phosphorus is larger and unstrained (7.12), the activation energy to reach the intermediate is greater and so reaction is slower. The model for the geometry of the inter-

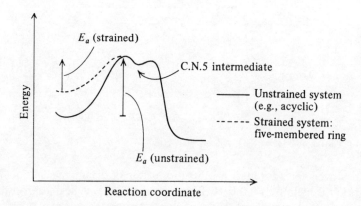

unstrained

(7.12)   Calculated and thermochemical ring strain, ~25 kJ mol⁻¹

mediate is that of stable isolable oxaphosphoranes (7.13). A schematic diagram of the reaction coordinates and related energetics, as outlined above, is presented in (7.14).

geometry of a model of C.N.5 intermediate

(7.13)   Geometries for cyclic molecules determined by X-ray diffraction

(7.14)   Reaction energetics – schematic

The hypothesis of a C.N.5 intermediate fits so many experimental results that it becomes increasingly more convincing. For example, the fast hydrolysis of a methyl phostonate with exclusive ring opening is readily understood in terms of an intermediate that cannot undergo ligand permutation, because this would place an electropositive ligand ($-CH_2$) axially–an energetically undesirable situation (7.15). The phostonate result also supports the idea that the entering group in the reaction comes in axially, and that the departing group must leave axially.

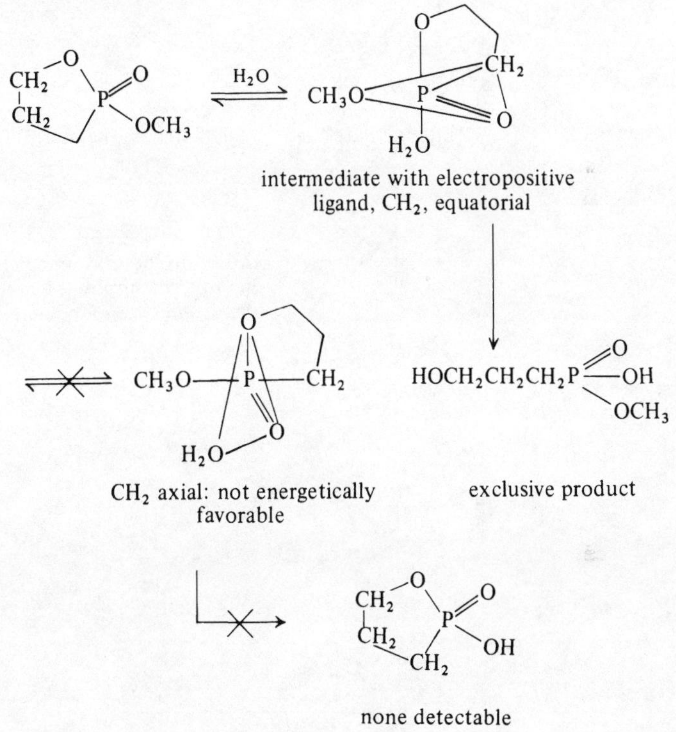

intermediate with electropositive ligand, CH_2, equatorial

CH_2 axial: not energetically favorable

exclusive product

none detectable

(7.15) Hydrolysis of a phostonate

Ligand interchange in a C.N.5 intermediate is frequently invoked in explaining many other kinetic and stereochemical results in displacement reactions without the wealth of supporting evidence and the many geometric models that are available for the cyclic phosphates. It is worth pointing out that C.N.5 intermediates should not always be accepted uncritically, without supporting evidence. As indicated earlier (Section 6.3), there do seem to be some displacement reactions at phosphorus that go directly without any necessity for specifying a discrete intermediate.

It is instructive to close this discussion by considering some theoretical results pertinent to reaction mechanisms. These arise from CNDO calculations of the (so far unobserved) reaction $F^- + F_3PO \rightleftharpoons F_4PO^-$, where F_4PO^- is a model for a C.N.5 intermediate. The attack of F^- on the F_3PO tetrahedron could occur on a face or along an edge, and in (7.16) the *energetics* of these various modes of attack are presented. Modes of attack in which the incom-

$$F^- + F_3PO \longrightarrow F_4PO^-$$

mode of attack	intermediate	calculated E_a (kJ mol^{-1}) relative
FFF face		O *kinetically* best intermediate; thermodynamically less stable than isomer with O^- equatorial
OFF face		63
FF edge		125
OF edge		288

(7.16)   Activation energies for $F^- + POF_3$ reaction

ing F^- is adjacent to the developing O^- are less favored because of increased electrostatic repulsion. Attack on a tetrahedral face is strongly favored over attack at an edge. The theoretical results are certainly not in conflict with the views developed earlier in this chapter on the nature of displacement reactions, but it must be added that theoretical chemistry has only just begun to examine problems involving third-period elements like phosphorus, because of their complexity. In the near future, with more powerful computers becoming available, theoretical chemistry should begin to be more informative and helpful with the chemistry of these heavier elements.

8 Phosphorus in its group

If we consider the chemistry of the range of elements in group VA, namely, N, P, As, Sb, Bi, then we are forced to conclude that it is the second-period element, nitrogen, that is anomalous. From this point of view, we can regard phosphorus as the lightest (and most abundant, and most fully studied) of the representative elements of the group. In fact, this perspective is a very helpful one in looking at the whole sweep of the main-group elements in the periodic table. The first- (H, He) and second-period elements are the ones with anomalous properties, whereas the heavier elements show more regularity. Compare carbon with silicon, germanium, and tin; or oxygen with sulfur, selenium, and tellurium. Because this work is devoted to phosphorus chemistry, however, we shall limit ourselves to comparisons within group V.

In (8.1) are listed the Pauling electronegativities and some properties of the trihydrides of N, P, As, and Sb. The sharp break in electronegativity comes

Element	Electronegativity	For EH$_3$			
		Boiling point (K)	ΔH_{fus} (kJ mol^{-1})	ΔH_{vap} (kJ mol^{-1})	ΔS_{vap} (J mol^{-1} K^{-1})
N	3.0	240	5.65	23.4	97.0
P	2.2	185	1.13	14.6	79.0
As	2.2	210	2.34	17.4	84.2
Sb	2.0	256	–	–	–

(8.1)   Group V element and hydride properties

between nitrogen and phosphorus, and so we expect that there will be a sharper differentiation at this point in the group than at any other. The electronegativity difference can be attributed to the relatively poor screening of the nucleus of nitrogen by the completed 1s shell and the incomplete 2s and 2p shells of electrons. In the other elements of the group there are filled sp shells underlying the valence shell, which screen the nucleus more effectively.

The properties of the hydrides shown in (8.1) are, of course, classic examples of the effects of hydrogen bonding. The highly electronegative nitrogen atom forms a very polar N—H bond which can hydrogen bond. However, the almost nonpolar P—H bond shows no tendency to hydrogen bond. (See the discussion of amine and phosphine basicities in Section 5.2.) Thus phosphine (PH_3) and arsine (AsH_3) have the properties of nonassociated hydrides as shown in their normal entropies of vaporization, low enthalpies of vaporization, and low boiling points. Ammonia, in contrast, has very high values of all these properties for a molecule of molecular mass 17, and displays similarly anomalous properties to its neighboring hydrogen-bonded analogs water (H_2O) and hydrogen fluoride (HF).

A second anomalous feature of the second-period elements that have lone-pair electrons is the weakness of their homopolar bonds. Thus bond energy terms for the group V elements lie in the order $N—N < P—P > As—As$ [see (8.3)]. The reason usually cited for this striking weakness in second-period elements, which is also manifested by oxygen as compared to sulfur, or fluorine as compared to chlorine, is the repulsive interaction of the relatively concentrated lone-pair electrons on adjacent nuclei. The more diffuse lone pairs on third- and fourth-period elements do not interact so strongly.

This weakness of the homopolar single bond in nitrogen leads to another striking contrast in chemistry. For phosphorus and arsenic, well-characterized cyclic catenated compounds containing homopolar bonds, the cyclopolyphosphines and arsines (8.2), are known. No analogous compounds are known for nitrogen.

Ring size	Compound	
	P	As
3	$(C_6H_5P)_3$	
4	$\begin{array}{ccc} CF_3 & & CF_3 \\ \diagdown & & \diagup \\ & P—P & \\ & \mid \quad \mid & \\ & P—P & \\ \diagup & & \diagdown \\ CF_3 & & CF_3 \end{array}$	
5	$(C_6H_5P)_5$; $(CF_3P)_5$	$(CH_3As)_5$
6	$(C_6H_5P)_6$	

(8.2)   Some cyclopolyphosphines and arsines

We have already discussed, in part, the next bonding consideration, which is the relative strength of π bonds in second-period elements as opposed to third- or fourth-period elements. The table of bond energy terms for homopolar bonds in (8.3) will serve as an illustration. The great strength of NN multiple bonds is attributable to efficient $2p_\pi$-$2p_\pi$ overlap, and the weakness of π bonds in the heavier elements is attributable to poor overlap of diffuse $3p$ or $4p$ orbitals.

N—N	P—P	As—As
170	230	About 170
N=N	P=P	As=As
420		
N≡N	P≡P	As≡As
940	480	380

(8.3) Bond energy terms $(kJ\ mol^{-1})$

Thus the chemistry of phosphorus and arsenic is full of polymeric singly bonded analogs of compounds that, when based on nitrogen, contain multiple bonds. An instructive elementary example is that of the forms of the elements at room temperature; nitrogen exists as gaseous N_2 containing NN triple bonds, whereas phosphorus (and also arsenic) exists as a solid containing three-coordinate singly bonded phosphorus atoms. In (8.4) a very simple, but instructive, calculation of the energetics of this phenomenon is presented, and

Process	Energy change (based on bond energy terms in $kJ\ mol^{-1}$)
$\frac{1}{2}N\equiv N \longrightarrow \diagdown N \diagup$ One triple bond, shared by two atoms Three single bonds, each shared by two atoms	+218 endothermic
$\frac{1}{2}P\equiv P \longrightarrow \diagdown P \diagup$	-104 exothermic

(8.4) Interconversion of allotropes of N and P

speaks for itself. Some further examples of compounds that have formally similar empirical formulas, but completely different structural formulas, because of this property of nitrogen multiple bonding are shown in (8.5).

The remaining comments to be made on phosphorus in its group are more conventional. As we progress from P to As to Sb to Bi we encounter more

Formula (E = N, P)	N structure	P structure
RE	RN$=\!=\!=$NR azo compound	$(RP)_n$ cyclopolyphosphine
E_2O_5		P_4O_{10} (see 6.35)
REO_2	 nitro compound	 polyphosphonate
EO_3^-	 nitrate	 polyphosphate

(8.5)   N and P: formal similarities and structural differences

and more metallic properties, and the chemistry of phosphorus is the most covalent of that of the heavier elements. We can compare, for instance, the gaseous volatile PF_3, with the crystalline ionic BiF_3. The chlorides show similar trends, with phosphorus trichloride undergoing very fast hydrolysis with production of hydrogen chloride and phosphorous acid, H_3PO_3. (The hydrolysis of NCl_3 is in the other sense, giving rise to NH_3 and $HOCl$, as a consequence of the high electronegativity of nitrogen.) The trichlorides of the other elements of the group are only slowly hydrolyzed, however, and eventually yield As_2O_3, $SbOCl$, and $BiOCl$, respectively, and these hydrolyses can be reversed by treating the end products with hydrogen chloride.

There is also a normal trend in the chemistry of the group oxides, in oxidation state V. The oxide of phosphorus is strongly acidic, that of arsenic is only weakly acidic, and antimony and bismuth(V) oxides are basic.

In closing this brief review of periodic properties, it is worth noting that the chemistry of nitrogen and phosphorus is much better developed than that of the heavier elements of this group. The chemistry of arsenic, after an early and fruitful period at the beginning of the twentieth century, was neglected for a long period, and the chemistry of antimony and bismuth has attracted relatively few investigators. There are certainly many interesting and worthwhile aspects of the chemistry of these elements that deserve fuller study.

9 Phosphorus reagents in chemistry

In this chapter selected examples of the use of phosphorus reagents in chemistry will be presented. The intention is to present areas where reagents based on phosphorus can carry out a unique transformation, or one that is especially useful. Some of the more obvious applications of phosphorus reagents (e.g., the use of P_4O_{10} as a desiccant, or the use of H_3PO_4 as a dehydrating agent for alcohols) will not be discussed, and the chapter, for reasons of length, will be selective. One important area that will not be covered in this chapter, because it was discussed earlier (Section 6.3.1), is the use of Wittig reagents in alkene synthesis.

9.1 Halogenation with P–halogen compounds

Both C.N.3 and C.N.5 compounds containing phosphorus–halogen bonds are commonly used as halogenating agents. For instance, in organic chemistry, phosphorus tribromide and phosphorus iodides (often made in situ by the direct combination of the elements) are widely used for the transformation $C-OH \longrightarrow C-X$ (where X = Br, I). It is worthwhile pointing out, because there still seems to be widespread misunderstanding on this point, that, generally speaking, phosphorus trichloride is *not* a useful reagent for the conversion of alcohols to alkyl chlorides. The usual course of the reaction between primary and secondary alcohols and PCl_3 is indicated in (9.1) and involves the production of a hydrogen phosphonate (dialkyl phosphite). Only with tertiary alcohols is there efficient conversion to alkyl chloride.

Other phosphorus–halogen compounds, however, can give excellent yields

$$3RR'CHOH + PCl_3 \longrightarrow (RR'CHO)_3P + 3HCl$$
not isolated
unless base
is present

$$(RR'CHO)_3P + HCl \longrightarrow (RR'CHO)_2P(:O)H + RR'CHCl$$
isolated product

(9.1) Normal reaction between primary and secondary alcohols and PCl_3

85

of halides from all classes of alcohols. For example, the triaryl phosphite di-halides, $(ArO)_3PX_2$ (where X = Cl, Br) convert alcohols to halides efficiently. Even sterically hindered alcohols, such as neopentyl alcohol, are reactive. With chiral alcohols, the stereochemical result is complete inversion; a plausible mechanism is shown in (9.2).

$$(C_6H_5O)_3P + Br_2 \longrightarrow (C_6H_5O)_3PBr_2$$

$$\xrightarrow{ROH} (C_6H_5O)_2(RO)\overset{+}{P}Br + C_6H_5OH$$

$$Br^- \qquad (C_6H_5O)_2P(:O)Br + RBr$$

(9.2) Mechanism of phosphite dihalide reaction with alcohols

An even more versatile reagent is the dihalide of triphenylphosphine, $(C_6H_5)_3PX_2$ (where X = Cl, Br), which not only converts alcohols to halides but also (like PCl_5) gives geminal dihalides from aldehydes and ketones. In addition, these reagents convert phenols into aryl halides directly, a very useful synthetic process (9.3).

$$(C_6H_5)_3PBr_2 \xrightarrow[80°C, 92\%]{(CH_3)_3CCH_2OH} (CH_3)_3CCH_2Br$$

$$(C_6H_5)_3PCl_2 + \text{[naphthol-OH]} \xrightarrow[79\%]{220°C} \text{[naphthyl-Cl]}$$

(9.3) Reactions of triphenylphosphine dihalides

9.2 Desulfurization

The formation of the strong P:S bond is the driving force for a number of preparatively useful reactions in which C.N.3 phosphorus compounds are used to remove sulfur from organic compounds. For example, the stereospecific conversion of a diol to an alkene can be effected using a trialkyl phosphite to desulfurize an intermediate thionocarbonate, as shown in (9.4) for the preparation of the unstable *trans*-cyclooctene in high optical purity.

$$\longrightarrow \text{[cyclooctene]} + CO_2 + S = P(OR)_3$$

(9.4) Phosphite desulfurization of a thionocarbonate

For monodesulfurization of disulfides, which can sometimes be a useful route to sulfides, the reagents of choice are the phosphorous triamides. Thus a cystine derivative can be cleanly converted to a lanthionine derivative (9.5). Kinetic and stereochemical studies of this reaction support a mechanism with a quasi-phosphonium intermediate as shown in (9.5) and with 100% inversion of configuration at *one* of the carbon centers in the original disulfide.

$$[CF_3CONHCH(CO_2CH_3)CH_2S]_2 + [(C_2H_5)_2N]_3P \longrightarrow$$

$$[CF_3CONHCH(CO_2CH_3)CH_2]_2S + [(C_2H_5)_2N]_3P:S$$

(9.5)   Desulfurization with phosphorous amides

9.3   Deoxygenation

The great affinity for oxygen of C.N.3 phosphorus compounds makes their utilization in deoxygenation reactions very understandable (although mechanistic details are often not clearly understood). Thus triphenylphosphine can be used under very mild conditions to deoxygenate ozonides (9.6). The corresponding carbonyl compounds are produced quantitatively, and this reaction has proved useful in natural product degradation studies.

(9.6)   Deoxygenation of ozonides

Deoxygenation of nitro and nitroso compounds by C.N.3 phosphorus compounds has proved to be a very versatile way of effecting reductive cyclization, and has been successfully exploited in the synthesis of a variety of heterocyclic systems (9.7). It seems likely that many of these reactions proceed via nitrenes or nitrenoid intermediates, although for many of them mechanisms have not yet been established.

(9.7)   Some heterocyclic syntheses from C.N.3 compounds and nitroaromatics

9.4   Dehalogenation

Halogenated phenols react rather slowly with triphenylphosphine under vigorous conditions, but in nevertheless useful yields, to lose halogen. Some typical reactions are indicated in (9.8).

(9.8)   Dehalogenation with triphenylphosphine

9.5   Oxiran synthesis from aldehydes

Phosphorous triamides react with aromatic aldehydes to produce good yields of oxirans (epoxides). The reaction is accelerated by electron-attracting substituents in the aldehyde and probably proceeds via a quasi-phosphonium intermediate as shown in (9.9). With some aldehydes of lower reactivity, the intermediate A (9.9) can be isolated and then, upon treatment with a more reactive aldehyde, it can yield an unsymmetrical oxiran.

$$2Ar—CHO + [(CH_3)_2N]_3P \longrightarrow Ar—CH\!\!-\!\!\underset{O}{\diagdown}\!\!-\!\!CH—Ar + [(CH_3)_2N]_3P{:}O$$

Examples:

93%

89%

$$(Me_2N)_3P{:} + \overset{\overset{\displaystyle H}{|}}{\underset{\underset{\displaystyle R}{|}}{C}}{=}O \longrightarrow (Me_2N)_3\overset{+}{P}—\overset{\overset{\displaystyle H}{|}}{\underset{\underset{\displaystyle R}{|}}{C}}—O^-$$

intermediate A $\overset{\displaystyle R'C{=}O}{\underset{\displaystyle H}{|}}$

$$\longrightarrow (Me_2N)_3\overset{+}{P}\!\!\underset{\overset{-}{O}\diagdown\!\!-CHR'}{\overset{\diagup CHR}{\diagdown}}\!\!O \longrightarrow (Me_2N)_3P({:}O) + RCH\!\!-\!\!\underset{O}{\diagdown}\!\!-\!\!CHR'$$

(9.9) Oxirans from aryl aldehydes

9.6 Ozonides of C.N.3 compounds

Many C.N.3 phosphorus compounds react with ozone to give rather unstable adducts that, upon warming, decompose to C.N.4 phosphoryl compounds and molecular oxygen. The O_2 is often produced (as would be expected from consideration of spin conservation) in the excited singlet state (probably $^1\Delta g$), and so the ozonides are often useful in effecting oxidations that are otherwise not easily carried out. The reactions shown in (9.10) are examples of the conversions possible with these reagents.

(−) terpinene   ascaridole

(9.10)   Oxidations with triphenyl phosphite ozonide

9.7   Oxophosphoranes in synthesis

Condensation between C.N.3 phosphorus compounds and 1,2-dicarbonyl compounds can lead to C.N.5 oxophosphoranes (Section 7.1), which are very versatile synthetic intermediates. Some synthetic schemes involving these intermediates, in which new carbon–carbon bonds are formed, are illustrated in (9.11).

(cont.)

(9.11)   Some syntheses from oxophosphoranes

9.8   Homogeneous catalyzed hydrogenation

Homogeneous hydrogenation of unsaturated organic compounds catalyzed by transition metal complexes has attracted much attention because of its synthetic potential and its mechanistic interest. The most useful transition metal catalysts contain phosphine ligands whose properties are crucial to the reactivity of the catalysts. This topic is a large and steadily expanding one. We shall limit our examination of it to a brief description of one of the most commonly used catalysts, $ClRh(PPh_3)_3$, often called Wilkinson's catalyst, and to a discussion of asymmetric hydrogenation with catalysts containing chiral phosphine ligands.

Alkenes are hydrogenated smoothly by hydrogen at room temperature and 1 atm pressure in presence of *tris*(triphenylphosphine)chlororhodium(I):

(9.12) $$C{=}C \ + H_2 \xrightarrow[20^\circ C,\ 1\ atm]{ClRh(PPh_3)_3} CH{-}CH$$

The advantages of this and other homogeneous catalysts include reactivity of similar level to the more commonly used heterogeneous catalysts (platinum or palladium, for example); more selectivity in hydrogenation; lower tendency to isomerize alkenes; and little tendency to hydrogenolyze cleavable groups. An example of selectivity is shown in (9.13).

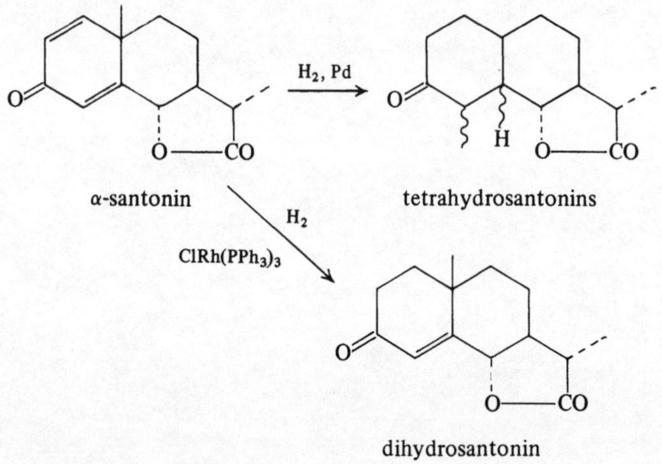

H₂, Pd
α-santonin
tetrahydrosantonins
H₂
ClRh(PPh₃)₃
dihydrosantonin

(9.13) Selective hydrogenation with $ClRh(PPh_3)_3$ as catalyst

The selectivity of homogeneous catalysts has led to the successful exploitation of their utility in asymmetric synthesis. In principle, hydrogenation of a suitable precursor on a chiral catalyst can lead to predominant production of a single enantiomer. Two examples of such a catalytic asymmetric synthesis are presented in (9.14), illustrating the use of phosphine ligands chiral either at carbon or at phosphorus.

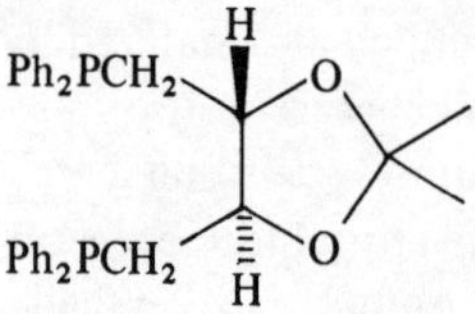

= DIOP, prepared optically active in several steps from tartaric acid

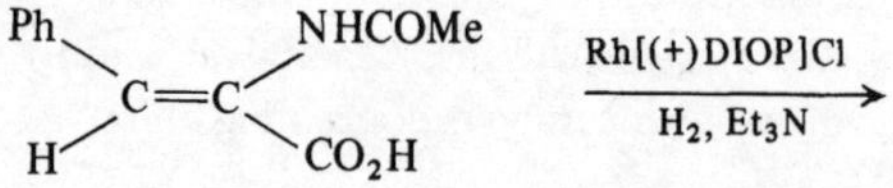

Rh[(+)DIOP]Cl
H₂, Et₃N

$PhCH_2CH(CO_2H)NHCOMe$
95% yield 72% enantiomeric excess of (L)-product

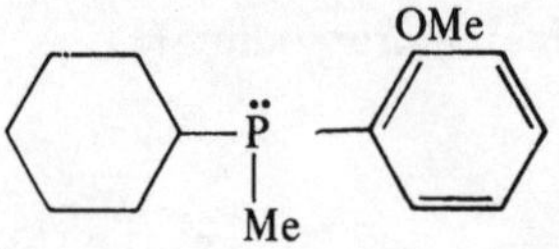

=P* prepared optically active

ClRh [(+)P*]₃
H₂

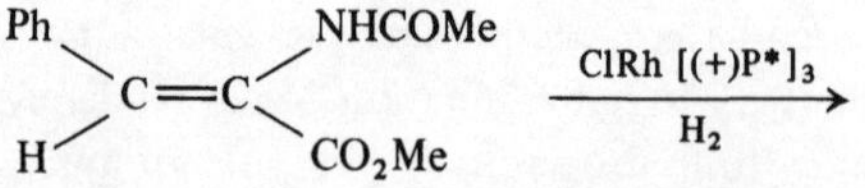

$PhCH_2CH(CO_2Me)NHCOMe$
90% yield 90% enantiomeric excess of (L)-product

(9.14) Asymmetric hydrogenation with chiral catalysts

Appendix I: Physical data

The structural data presented are from the following sources:

D. E. C. Corbridge, in *Topics in Phosphorus Chemistry*, Vol. 3 (1966) M. Grayson and E. J. Griffiths, eds. Interscience, New York. Structural chemistry of phosphorus compounds; 300-page catalog of information up to 1965.

D. E. C. Corbridge, *Structural Chemistry of Phosphorus*, Elsevier, Amsterdam, 1974. Exhaustive; literature coverage to early 1973.

L. S. Khaikin and L. V. Vilkov, *Russian Chem. Rev. 40*, 1014–29 (1971). Molecular structure of acyclic organophosphorus compounds. Very handy brief review, with many summary tables; literature to 1971.

Selected bond lengths

Bond type	Compound and phase	Bond length (pm)	Bond length (Å)
P—H	$PH_3(g)$	144	1.44
P—D	$PD_3(g)$	152	1.52
P—F	$PF_3(g)$	154	1.54
P—Cl	$PCl_3(g)$	204	2.04
P—Br	$PBr_3(g)$	218	2.18
P—I	$PI_3(c)$	247	2.47
P—C	$P(CH_3)_3(g)$	184	1.84
P—C	$P(C_6H_5)_3(c)$	183	1.83
P=C	$CH_2{=}P(C_6H_5)_3(c)$	166	1.66
P≡C	$HC{\equiv}P(g)$	154	1.54
P=O	$Cl_3P{=}O(g)$	145	1.45
P=O	$(CH_3)_3P{=}O(c)$	148	1.48
P—O—R	$(C_6H_5CH_2O)_2PO_2H(c)$	156	1.56
P=S	$(C_2H_5)_2P{=}S(c)$	186	1.86
P—P	$P_4(c)$	221	2.21
P—P	$(CF_3P)_5(c)$	222	2.22
P—N	$(C_6H_5)_2P({:}O)N(CH_3)_2(c)$	167	1.67
P=N	$(Cl_2P{=}N)_3(c)$	159	1.59
P≡N	$P{\equiv}N(g)$	149	1.49

Selected bond energies

The user is cautioned that the values presented here may contain substantial errors. The thermochemistry of phosphorus is still undeveloped. In compounds where more than one type of bond is present, experimental heats of atomization have been divided according to some arbitrary scheme. For example, the value of the bond energy of the P=O bond in $F_3P=O$ is derived by making the (dubious) assumption that the bond energy of the PF bonds in $F_3P=O$ is the same as that found in PF_3.

Bond type	Molecule	$\bar{D}$ (kJ mol^{-1})	$\bar{D}$ (kcal mol^{-1})
P—H	PH_3	322	77
P—F	PF_3	489	117
P—Cl	PCl_3	318	76
P—Br	PBr_3	260	62
P—I	PI_3	184	44
P—P	P_4	209	50
P—O(—P)	P_4O_6	360	86
P—O(—R)	$P(OCH_3)_3$	380	91
P=O	P_4O_{10}	576	138
P=O	$(CH_3O)_3P=O$	625	150
P=O	$F_3P=O$	543	130
P=O	$(CH_3)_3P=O$	580	139
P=S	$Cl_3P=S$	293	70
P=S	$(C_2H_5O)_3P=S$	381	91
P—C	$(CH_3)_3P$	272	65
P—C	$(C_6H_5)_3P$	297	71
P≡C	CP	665	159
P—N	$[(CH_3)_2N]_3P$	293	70
P=N	$(CH_3)_3P=NC_2H_5$	405	97
P=N	$(Cl_2P=N)_3$	334	80
P≡N	PN	685	164

Appendix II: Spectroscopic data

The following tables are included as a guide to ranges of normal spectroscopic values. For more detailed information, consult the Suggestions for Further Reading for Chapter 3.

Table 1. *Infrared spectroscopic data*

Group	Group range (in cm^{-1}) and usual intensity (s = strong, m = medium, w = weak)
P—H	2440-2280 (m)
P=O	Much affected by compound type and association (see text, Section 3.1); typical ranges:
	(Alkyl-O)$_3$P=O 1290-1255 (s)
	(Aryl-O)$_3$P=O 1315-1290 (s)
	(Aryl)$_3$P=O 1145-1095 (s)
	(Alkyl)$_2$P(=O)OH 1205-1135 (s)
P=S	Not a very useful correlation; usually two bands:
	805-655 (w)
	730-550 (w)
P—O alkyl	1055-990 (s)
P—CH$_3$	1320-1280 (m)
P—Cl	580-440 (s)
P—F	885-810 (s)

Table 2. ^{31}P chemical shifts

All expressed in the scale in parts per million, with 85% H_3PO_4 (ext.) as zero. Low-field shifts are given a *positive* sign. (Note that this is the reverse of the earlier convention used in various compilations, but is the IUPAC recommended convention.)

Compound	δ	Compound	δ
C.N.3			
CH_3PF_2	+245	$(C_2H_5)_3P$	-20
PBr_3	+227		
		$(C_2H_5)_2PH$	-57
PCl_3	+220		
		$C_2H_5PH_2$	-128
PI_3	+178		
		PH_3	-241
PF_3	+97		
		KPH_2	-255
$(CH_3O)_2PCl$	+169		
		CH_2—CH_2 $\diagdown PH \diagup$	-341
$(CH_3)_2PCl$	+96		
		P_4	-462
C.N.4			
$(C_2H_5)_3PS$	+54		
F_3PS	+32	H_3PO_4	0
Cl_3PS	+30	F_3PO	-93
$CH_3P(O)F_2$	+27	Br_3PO	-102
$(CH_3)_4P^+Br^-$	+25	Br_3PS	-112
$(CH_3O)_2P(:O)H$	+10		
$(CH_3O)_2P(:O)Cl$	+8		
Cl_3PO	+3		
C.N.5 and 6			
PH_2F_3	-24	PCl_5	-80
		PBr_5	-101
$(CH_3O)_3P \overset{\displaystyle O\diagdown}{\underset{\displaystyle O\diagup}{}} \overset{\displaystyle CH_3}{\underset{\displaystyle CH_3}{}}$	-50		
		PF_6^-	-145
PF_5	-80	PCl_6^-	-295

Coupling constants

All coupling constants are given in hertz. Where signs are known they are given explicitly. Values without signs are of uncertain sign.

Table 3. $^1J_{PH}$

Compound	$^1J_{PH}$ (Hz)	Compound	$^1J_{PH}$ (Hz)
HPF_4	1075		
		$(C_6H_5)_2P(:O)H$	486
$CF_3PF_4H^-$	943		
		$C_6H_5PH_2$	197
$HP(:O)F_2$	878	$(CH_3)_2PH$	+192
H_2PF_3	841	CH_3PH_2	187
$(C_2H_5O)_2P(:O)H$	688	PH_3	+182
PH_4^+	+547	$\underset{\diagdown\;PH\;\diagup}{CH_2-CH_2}$	155
$CH_3PH_3^+$	527	PH_2^-	139

Table 4. $^1J_{PF}$

Compound	$^1J_{PF}$ (Hz)	Compound	$^1J_{PF}$ (Hz)
PF_3	-1441	PF_5	938
PCl_2F	1320	CH_3PF_3H	968 (F equatorial)
			805 (F axial)
$C_6H_5P(:O)F_2$	1105	$(CF_3)_3PF_2$	881
F_3PO	1055	PF_6^-	706
$(CF_3)_2PF$	-1013	$(CH_3)_3PF_2$	541

Table 5. $^1J_{P^{13}C}$

Compound	$^1J_{P^{13}C}$ (Hz)	Compound	$^1J_{P^{13}C}$ (Hz)
$CH_3P(O)Cl_2$	104	HCP	54
$(C_6H_5)_4P^+$	88	$(CH_3)_3P$	-14
$(CH_3)_4P^+$	+56	CH_3PH_2	9

Table 6. $^2J_{PCH}$

Compound	$^2J_{PCH}$ (Hz)	Compound	$^2J_{PCH}$ (Hz)
CH_3PCl_2	17.6	$(CH_3)_3PO$	-13.4
CH_3PF_2	10.2	$(CH_3)_4P^+$	-14.6
CH_3PH_2	+3.9	$CH_3PH_3^+$	-17.6

Table 7. *Other representative ^2J values*

Compound	^2J type	Value (Hz)
CH_3PH_2	HPH	-12
HPF_2	HPF	+41
CF_3PCl_4	PCF	154
$(CF_3)_3PO$	PCF	113
$(CF_3)_3P$	PCF	86
CF_3PH_2	PCF	+48

Suggestions for further reading

References are listed by chapter and section. Citations in bold type refer to general references in the Bibliography that have been assigned short titles. For full references, see the Bibliography.

Chapter 1
1.1 For discovery and early history of phosphorus, see M. E. Weeks and H. M. Leicester, *Discovery of the Elements,* Journal of Chemical Education Publishers, Easton, Pa., 1968, pp. 110–30.

1.2 *Chemistry in the Economy,* American Chemical Society, Washington D.C., 1973, chaps. 2, 8, 10, 11.

A. Grangow, *Accts. Chem. Res. 11,* 177 (1978), discusses flame retardation by phosphorus compounds, an important application of phosphorus chemistry.

Mellor, chaps. 1 and 2.

Van Wazer, vol. 2.

1.3 E. J. Griffith, ed., *Environmental Phosphorus Handbook,* Wiley, New York, 1973. A comprehensive multiauthored compendium (718 pages) of the part phosphorus plays in the environment. An excellent source book, with literature coverage to 1971.

G. E. G. Mattingly and O. Talibudeen, "Progress in the Chemistry of Fertilizer and Soil Phosphorus," in Topics *4* (1967). 130 pages.

Chapter 2
2.1 U. Schmidt, "Formation, Detection and Reaction of Phosphinidenes (phosphanediyls)," *Angew. Chem. Int. Ed. 14,* 523–9 (1975). Intermediates of type RP.

2.2 K. Dimroth, *Topics in Current Chemistry 38,* Springer Verlag, Berlin, 1973. Phosphorus–carbon double bonds. 140 pages, literature to 1972. Covers the work of Dimroth's group and others on phosphabenzenes and phosphacyanins.

P. Jutzi, "New Element–Carbon (*p–p*) Bonds," *Angew. Chem. Int. Ed. 14,* 232–45 (1975). Includes PC double bonds. Literature to 1974.

2.3–2.5 See references for Chapters 5, 6, and 7.

2.6 D. Hellwinkel, in **Kosolapoff and Maier** *3.*

Chapter 3
3.1 **Halmann.**

D. E. C. Corbridge, in Topics *6,* 235–365 (1969). Surveys vibrational spectroscopy of both inorganic and organic compounds. Detailed, good bibliography, literature to 1967.

Spec. Props.

L. C. Thomas, *Interpretation of Infrared Spectra of Organophosphorus Compounds*, Heyden, London, 1974. Includes all major functional group correlations and many spectra with worked-out interpretations. Very useful for organophosphorus research workers. 270 pages, literature to 1970.

3.2 J. F. Brazier, D. Houalla, M. Koenig, and R. Wolf, in Topics *8*, 99–192, have an extensive discussion of nmr parameters of the proton directly bonded to phosphorus, with tabulation of many values of coupling constants and chemical shifts. Literature examined to 1974.

J. W. Emsley, J. Feeney, and L. H. Sutcliffe, eds., *Progress in nmr Spectroscopy*, Pergamon, Oxford. A. Finer and R. Harris, in vol. 6 (1971), give a full review of phosphorus-phosphorus coupling constants. 50 pages, literature to 1970.

Halmann.

J. K. Kochi, ed., *Free Radicals*, vols. 1 and 2, Wiley-Interscience, New York, 1973. W. G. Bentrude, chapter 22 (vol. 2) on phosphorus-containing radicals. Literature to 1971.

Kosolapoff and Maier for specific compounds.

J. W. Emsley, J. Feeney, and L. H. Sutcliffe, eds., *Progress in nmr Spectroscopy*, Pergamon, Oxford. G. Mavel in vol. 1 (1966) covers studies of phosphorus compounds using magnetic resonance of nuclei other than ^{31}P. 120 pages. Covers the topic comprehensively from its beginnings to the end of 1965.

E. F. Mooney, ed., *Annual Reports on nmr spectroscopy B*, Academic Press, London, G. Mavel, vol. 5B (1973), *N.m.r. Studies of Phosphorus Compounds, 1965-1969.* Comprehensive collection; 400 pages of chemical shifts and coupling constants, with interpretative discussion.

P. Schipper, E. H. J. M. Jansen, and H. M. Buck, in Topics *9*, 407–504, collect and discuss the esr of phosphorus compounds. Literature up to mid-1974 is covered.

Specialist Periodical Reports. Electron spin resonance, The Chemical Society, London. Annual review. Volume 1 covers 1972 literature, vol. 2, 1973, etc.

Spec. Props.

Topics *5*, is devoted totally to ^{31}P nmr, with literature coverage to 1967, and articles as follow: M. Crutchfield, C. H. Dungan, and J. Van Wazer, "Measurement and Interpretation of High-Resolution ^{31}P nmr" (70 pages); J. H. Letcher and J. Van Wazer, "Quantum Mechanical Theory of ^{31}P Chemical Shifts" (90 pages); J. Van Wazer and J. H. Letcher, "Interpretation of Experimental ^{31}P Chemical Shifts and Some Remarks Concerning Coupling Constants" (60 pages); V. Mark, C. H. Dungan, M. Crutchfield, and J. Van Wazer, "Compilation of ^{31}P nmr Data" (200 pages; easily the most valuable part of the book; some coupling constants are also included).

3.3 I. Granoth, in Topics *8*, 41–98, reviews mass spectra of organophosphorus compounds, with literature coverage to 1973.

M. R. Litzow and T. R. Spalding, *Mass Spectrometry of Inorganic and Organometallic Compounds*, Elsevier, Amsterdam, 1973. Chapter 8 has 120 pages on group V compounds, the majority of it on phosphorus compounds. Results and interpretation are presented critically; literature coverage to early 1971.

Specialist Periodical Reports. Mass Spectrometry, The Chemical Society, London; vol. 1, June 1968–June 1970; vol. 2, June 1970–June 1972; vol. 3, June 1972–June 1974.

3.4 H. Bock and B. G. Ramsey, *Angew. Chem. Int. Ed. 12*, 734 (1973), examine photoelectron spectra of nonmetal compounds and their interpretation by MO methods. Literature through 1972.

V. V. Zverev and Y. T. Kitaev, *Russian Chem. Rev. 46*, 791 (1977), review application of photoelectron spectroscopy to phosphorus compounds, with literature coverage through 1976.

Chapter 4

4.1 H. A. Bent, *Chem. Rev. 61*, 275 (1961). A key account of simple hybridization theory.

Hudson, chap. 1.

E. N. Tsvetkov and M. I. Kabachnik, *Russian Chem. Rev. 40*, 97–125 (1971). Conjugation in phosphorus(III) compounds. Literature to 1970. Very useful collection of data and interpretations on a controversial topic.

4.3 K. A. R. Mitchell, *Chem. Rev. 69*, 157 (1969). Full discussion of the question of *d*-orbital involvement in bonding.

H. R. Allcock, *Phosphorus-Nitrogen Compounds*, Academic Press, New York, 1972. 450 pages, mostly on cyclophosphazenes and their polymers. Covers literature to mid-1971.

A. J. Ashe, *Accts. Chem. Res. 11*, 153 (1978), has an authoritative review on the group V heterobenzenes, with much discussion of aromaticity in phosphobenzene derivatives.

C. A. Coulson, *Nature 221*, 1106 (1969). Theoretical chemical results pertaining to *d*-orbital involvement.

Hudson, chap. 1.

S. S. Krishnamirthy, A. C. San, and M. Woods, *Adv. Inorg. Chem. Radiochem. 21*, 41 (1978), update the subject of cyclophosphazenes to mid-1977 in a 75-page review; especially strong on physical methods.

H. Kwart and K. King, *d-Orbital Involvement in the Organo-Chemistry of Silicon, Phosphorus and Sulfur*, Springer-Verlag, New York, 1977.

See also references for Section 2.2.

Chapter 5

5.1 Houben-Weyl.

Kosolapoff and Maier.

The following are also of interest for compounds of the specified types.

F. A. Cotton, ed., *Progress in Inorganic Chemistry*, Interscience, New York. L. Maier, in vol. 5 (1963), gives a comprehensive survey of the preparations and properties of primary, secondary, and tertiary phosphines (180 pages, literature to 1962).

W. L. Jolly, ed., *Preparative Inorganic Reactions*, Interscience, New York. W. L. Jolly and A. D. Norman, in vol. 4, (1968), chap. 1, "Hydrides of Groups IV and V," discuss preparation of unsubstituted phosphanes (literature to 1967). E. Fluck, in vol. 5 (1968), describes synthesis of compounds containing P–P bonds (literature to 1967).

M. Stacey, J. C. Tatlow, and A. G. Sharpe, eds., *Advances in Fluorine Chemistry*, Butterworths, London. R. D. Kermitt and D. W. A. Sharpe, in vol. 4 (1965), give brief notes (10 pages) on preparation and properties of phosphorus fluorides (literature to 1963). R. Schmutzler, in vol. 5 (1965), describes in detail (250 pages) compounds containing PF bonds with tables of properties (literature to 1964).

5.2 N. L. Allinger and E. L. Eliel, eds., *Topics in Stereochemistry*, Interscience, New York. M. J. Gallagher and I. D. Jenkins, in vol. 3 (1968), review stereochemical aspects of phosphorus chemistry (100 pages, with literature coverage to 1967). J. B. Lambert, in vol. 6 (1971), discusses pyramidal inversion (literature to 1970).

A. Rauk, L. C. Allen, and K. Mislow, *Angew. Chem. Int. Ed. 9*, 400 (1970), also discuss pyramidal inversion, with special attention to phosphorus compounds (literature to 1969).

5.3 E. Arnett, *Accts. Chem. Res. 6*, 404 (1973), describes and interprets results on gas-phase basicities, covering literature to 1972.

Hudson, chaps. 2, 4.

5.4 H. J. Emeléus and A. G. Sharpe, eds., *Advances in Inorganic Chemistry and Radiochemistry*, Academic Press, New York. J. F. Nixon, in vol. 13 (1970), describes recent progress in the chemistry of fluorophosphines with much attention to their ligand properties (100 pages, literature to 1969).

C. A. McAuliffe, ed., *Transition Metal Complexes of Phosphorus, Arsenic, and Antimony Ligands*, Macmillan, London, 1973. Very full collection of compounds made, and detailed discussions of the arguments on metal-ligand bonding. Literature to 1971.

C. A. McAuliffe and W. Levason, *Phosphine, Arsine and Stibine Complexes of the Transition Elements*, Elsevier, Amsterdam, 1979, brings the subject up to 1977, in 540 pages.

R. Mason and D. W. Meek, *Angew. Chem. Int. Ed. 17*, 183 (1978), discuss the properties that make tertiary phosphine ligands so useful in coordination and organometallic chemistry.

O. Stelzer, in Topics *9*, 1–230, presents a book-length review and tabulation of transition metal complexes of phosphorus ligands, in which the focus is on 3-C.N. phosphorus donors, with selected literature coverage to 1973.

C. A. Tolman, *Chem. Rev. 77*, 313 (1977), discusses the role steric effects play in the ligand properties of 3-C.N. phosphorus compounds. Literature to 1976.

5.5 Organo-P.

5.6 J. C. Lockhart, *Redistribution Reactions*, Academic Press, New York, 1970. Chapter 8 is on reactions in group V with literature coverage to 1969.

5.7 J. I. G. Cadogan, *Quart. Rev. 16*, 208 (1962), reviews oxidation of tervalent organic compounds of phosphorus (300 pages), covering literature to 1961.

Houben-Weyl.

Chapter 6

6.1 L. Almasi, *Les Composés Thiophosphororganiques*, Masson, Paris, 1976. A 350-page book on preparations, structure, bonding, and dynamics of organophosphorus compounds containing the P:S group. Literature to 1974.

H. Hoffman and M. Becke-Goehring, in *Topics 8*, 193–272, give a detailed review including preparations, of the inorganic phosphorus sulfides, with literature covered to 1973.

R. G. Harvey and E. R. De Sombre, "Michaelis-Arbuzov and Related Reactions," *Topics 1* (1964). 55 pages, literature reviewed to 1963.

Houben-Weyl.
Kosolapoff and Maier.
L. Maier, "Preparation and Properties of Primary, Secondary, and Tertiary Phosphine Sulfides and Related Compounds," Topics 2 (1965). 90 pages, literature reviewed to 1964.
B. Miller, "Reactions Between Trivalent Phosphorus Derivatives and Positive Halogen Sources," Topics 2 (1965). 50 pages, literature reviewed to 1964.
C. V. Shen and C. F. Callis, "Orthophosphoric Acids and Orthophosphates," in W. L. Jolly ed., *Preparative Inorganic Reactions 2*, Interscience, New York, 1965. 20 pages, literature 1963.
6.2 W. McEwen, "Stereochemistry of Reactions of Organophosphorus Compounds, Topics 2 (1965) 42 pages, literature reviewed to 1964.
(See also Allinger and Eliel, cited under Section 5.2.)
6.3 V. E. Belskii, *Russian Chem. Rev. 46*, 828 (1977), surveys kinetics of hydrolyses of phosphate esters, with literature covered to 1976.
H. J. Bestmann, *Angew. Chem. Int. Ed. 16*, 349 (1977), discusses the synthetic uses of phosphacumulene and phosphoallene ylids, $:P:C:C:X$ (X = O, S, NR, CR_2).
G. M. Blackburn and J. S. Cohen, "Chemical Oxidative Phosphorylation," Topics 6 (1969). 50 pages, literature to 1968.
D. M. Brown, "Phosphorylation," in R. A. Raphael, E. C. Taylor, and H. Wynberg, eds., *Advances in Organic Chemistry 3*, Interscience, New York, 1963. 80 pages, literature to 1961. Emphasizes techniques and reagents.
V. A. Clark and D. W. Hutchinson, "Phosphoryl Transfer," in J. Cook and W. Carruthers, eds., *Progress in Organic Chemistry 3*, Butterworths, London, 1968. 40 pages, literature to 1967. Logical classification of transfer mechanisms.
W. Foerst, ed., *Newer Methods of Preparative Organic Chemistry 5*, Academic Press, New York, 1968. H. J. Bestmann has a 60-page review on new reactions of alkylidenephosphoranes and their preparative uses, with literature coverage to 1966.
Hudson.
A. W. Johnson, *Ylid Chemistry*, Academic Press, New York, 1966. Full discussion of phosphorus ylids with preparative details, uses, and mechanistic interpretation. Literature coverage to 1965.
Kirby and Warren.
Kosolapoff and Maier.
Organo-P.
R. K. Osterheld, "Nonenzymic Hydrolysis at Phosphate Tetrahedra," Topics 7 (1972). 150 pages, literature to 1971.
W. S. Wadsworth, *Org. Reactions 25*, 73 (1977), surveys comprehensively (180 pages) the synthetic applications of the Wittig-like reagents derived from phosphoryl-stabilized anions, $R_2P(:O)CHR' \longleftrightarrow R_2P(-O^-):CHR'$.
6.4 M. Bermann, "The Phosphazotrihalides," in *Advances in Inorganic Chemistry and Radiochemistry* H. J. Emeléus and A. G. Sharpe, eds., Academic Press, London. *14*, 1972. 120 pages, literature to 1971.
M. Bermann, "Compilation of Physical Data of Phosphazo Trihalides," Topics 7 (1972). (Includes catenated compounds with $-N=PX_3$ end group). 70 pages, literature to 1971.
J. D. Curry and D. A. Nicholson, "Oligophosphonates," Topics, 7 (1972). 65 pages, literature to 1971.
S. Y. Kalliney, "Cyclopolyphosphates," Topics 7 (1972). 55 pages, literature to 1971.

L. B. Kubasova, *Russian Chem. Rev. 40,* 1-12 (1971). Polyphosphoric acids. Literature to 1970.

S. Ohashi, "Condensed Phosphates Containing Other Oxo Acid Anions," Topics *1* (1964). 50 pages literature to 1963.

C. Y. Shen and D. R. Dynoff, "Condensed Phosphoric Acids and Condensed Phosphates," in W. L. Jolly, ed., *Preparative Inorganic Reactions 5,* Interscience, New York, 1968. 65 pages, literature to 1966.

E. Thilo, "Condensed Phosphates and Arsenates," in *Advances in Inorganic Chemistry and Radiochemistry* H. J. Eméleus and A. G. Sharpe, eds., Academic Press, London. *4,* 1962. 75 pages, literature to 1961.

Chapter 7

7.1 B. A. Arbuzov and N. A. Polezhaeva, *Russian Chem. Rev. 43,* 414-33 (1974). Cyclic oxyphosphoranes. Literature to 1973.

V. Gutmann, ed., *Halogen Chemistry,* Academic Press, New York, 1967. Volume 2 has a review of fluorophosphoranes by R. Schmutzler (80 pages, literature to 1966).

Kosolapoff and Maier.

F. Ramirez, *Bull. Soc. Chim. France,* 2443-50 (1966). Recent developments in the chemistry of oxyphosphoranes. Literature to 1965.

F. Ramirez, *Bull. Soc. Chim. France,* 3491-3519 (1970). Synthesis and dynamic stereochemistry of oxyphosphoranes. Literature to 1969.

7.2, 7.3 P. Gillespie, P. Hoffmann, H. Klusacek, D. Marquarding, S. Pfohl, F. Ramirez, E. A. Tsolis, and I. Ugi, *Angew. Chem. Int. Ed. 10,* 687-715 (1971). Nonrigid molecular skeletons. Berry pseudorotation and turnstile rotation. Literature to 1971.

P. Gillespie, F. Ramirez, I. Ugi, and D. Marquarding. *Angew. Chem. Int. Ed. 12,* 91-119 (1973). Displacement reactions of phosphorus(V) compounds and their pentacoordinate intermediates. Literature to 1972.

F. Ramirez and I. Ugi, "Turnstile Rearrangement and Pseudorotation in the Permutational Isomerization of Pentavalent Phosphorus Compounds," in V. Gold, ed., *Advances in Physical Organic Chemistry 9,* Academic Press, London, 1971. 100 pages, literature to 1970.

I. Ugi and F. Ramirez, *Chem. Brit. 8,* 198-206 (1972). Stereochemistry of five-coordinate phosphorus. Literature to 1971.

Each of the above reviews does have a somewhat different theme, but looking at them as a whole one is struck by a good deal of repetition–a case of academic overkill. Gillespie et al. (1973) is the most comprehensive, so it should be consulted first. The authors devised the turnstile rotation mechanism and are (understandably) enthusiastic about it. In addition, see also the following.

R. R. Holmes, *Accts. Chem. Res. 12,* 257 (1979), presents an authoritative overview of structures and dynamics in C.N.5. phosphorus compounds containing rings, comparing phosphorus with other main-group elements.

R. Luckenback, *Dynamic Stereochemistry of Pentaco-ordinated Phosphorus and Related Elements,* Georg. Thieme, Stuttgart, 1973. Covers the published work up to mid-1972 lucidly in 260 pages. Very thorough coverage.

Chapter 8

R. T. Sanderson, *Inorganic Chemistry,* Van Nostrand Reinhold, New York, 1967. The most consistent and lucid examination of the periodic properties of the elements.

Chapter 9

J. I. G. Cadogan and R. K. Mackie, *Chem. Soc. Rev. 3*, 87–137 (1974). Tervalent phosphorus compounds in organic synthesis. Very thorough and broader than the title, because it includes some mention of reagents derived from the title compounds (e.g., phosphite ozonides). Literature to 1973.

L. Horner, *Fortsc. Chem. Forsch. 7*, 1–61 (1966). Preparative phosphorus chemistry. An admirably full, yet concise, review. Literature to 1965.

F. J. McQuillin, *Homogeneous Hydrogenation in Organic Chemistry*, D. Reidel Publishing Co., Dordrecht-Holland, 1976, gives in an admirably clear and concise (130 page) monograph a full review of the subject with literature coverage to 1975.

T. Mukaiyama and H. Takei, in Topics *8*, 587–646 1976, review the preparative and mechanistic aspects of the reactions of disulfides with trivalent phosphorus compounds, with literature coverage to 1974.

F. Ramirez, *Synthesis 90* (1974), describes very thoroughly the use of oxyphosphoranes in organic synthesis, and includes typical experimental conditions. Literature through 1973.

Bibliography

In these works will be found material relevant to several sections of the present book. The scope of these general works will be described. For ease of cross-referencing, some works in this section have been assigned short reference titles in bold print (given in parentheses at the end of the description).

C. C. Addison, ed. *Specialist Periodical Reports. Inorganic Chemistry of the Main-Group Elements,* The Chemical Society, London. A companion work to **Trippett,** in which a chapter is devoted to group V, most of which is on phosphorus chemistry. Volume 1 covers literature from July 1971 to September 1972; vol. 2 continues to September 1973, vol. 3 to September 1974; and so on.

J. Bailar, H. Eméleus, R. Nyholm, and A. Trotman-Dickenson, eds., *Comprehensive Inorganic Chemistry,* vol. 2, Pergamon Press, Oxford, 1973. Chapter 20, by A. D. F. Toy, is on phosphorus. More than 150 pages on strictly inorganic chemistry. Concise, comprehensive, but not well indexed. Literature coverage to 1967.

D. E. C. Corbridge, *Structural Chemistry of Phosphorus,* Elsevier, Amsterdam, 1974. Comprehensive collection of X-ray diffraction results with some treatment of electron diffraction and vibrational spectroscopy. Literature coverage to early 1973 is quite comprehensive and detailed, although the indexes are not as full as they might be. The first place to turn for structural data.

D. E. C. Corbridge, *Phosphorus,* Elsevier, Amsterdam, 1978. A comprehensive text (480 pages). In many ways complementary to **Emsley and Hall.** Strong on simple inorganic systems, preparations, and reactions. Rather sparse on bonding, stereochemistry, and applications of modern physical methods. Literature to 1977 is almost exclusively secondary and tertiary.

J. Emsley and D. Hall, *The Chemistry of Phosphorus,* Wiley, New York, 1976. A modern, extended look at all phases ("environmental, organic, inorganic, biochemical and spectroscopic aspects" as the book's subtitle promises) of phosphorus chemistry. The book is clearly written and is an excellent, quite thorough, introduction to the subject. Literature coverage is up to early 1974, is complete, and includes both primary and secondary literature. The only problem with this book is that its length (560 pages) may prove daunting to a novice in the field, but it is logically arranged and well indexed.

Gmelin's Handbuch der Anorganischen Chemie, Phosphor, Part B, System No.

16, Verlag Chemie, Weinheim, 1964. The usual high standard of totally in-
clusive survey of a limited range of topics. In 450 pages the technology of
phosphorus and H_3PO_4 production; the properties (physical and chemical)
of the element; and the detection and determination of phosphorus are
covered. Literature is to the end of 1960. *Part C*, 1965, has 642 pages on
the inorganic compounds of phosphorus combined with H, O, N, halogens,
S, Se, Te, B, and C (including, somewhat oddly, perfluorocarbon derivatives).
Literature is again covered to the end of 1960.

M. Grayson and E. J. Griffith, eds., *Topics in Phosphorus Chemistry*, Interscience, New York. A continuing series of generally excellent reviews, nine
volumes to date. **(Topics)**

M. Halmann, ed., *Analytical Chemistry of Phosphorus Compounds*, Wiley-
Interscience, New York, 1972. A comprehensive survey (830 pages) of
methods of separation and identification of all classes of phosphorus compounds. Contains classical wet methods, specific group tests, and spectroscopic techniques. Literature coverage to 1971. **(Halmann)**

Houben-Weyl: See Muller.

R. F. Hudson, *Structure and Mechanism in Organophosphorus Chemistry*, Academic Press, London, 1965. A well-organized and readable account of the
principles of the subject, containing many insights. Very useful as a starting
point though, inevitably, outdated in details. Literature to 1963. **(Hudson)**

A. J. Kirby and S. G. Warren, *The Organic Chemistry of Phosphorus*, Elsevier,
Amsterdam, 1967. A volume in a series on organic reaction mechanisms,
this book suffers somewhat from an overambitious attempt to classify mechanisms in organophosphorus chemistry when few mechanistic details were
available. Useful, nevertheless, for its assembling of reaction types; 400
pages, with literature coverage to 1965. **(Kirby and Warren)**

F. Korte, ed., *Methodicum Chemicum*, vol. 7, Academic Press, New York,
1978, contains discussions of the chemistry of, and analytical methods for a
wide range of inorganic and organic phosphorus compounds (200 pages, literature to 1976).

G. M. Kosolapoff, *Organophosphorus Compounds*, Wiley, New York, 1950.
The early "Bible" of organophosphorus chemistry covers the whole subject
in 376 pages, interpreting the topic liberally in that compounds with no P–C
linkages (e.g., esters and amides of phosphorus acids) are included. Strong
on synthesis and tables of compounds with properties; weaker on general
chemistry. Literature commendably up to date (to 1949), and the book is
still useful.

G. M. Kosolapoff and L. Maier, eds., *Organophosphorus Compounds*, Wiley
New York. A laudable updating of the original Kosolapoff volume (above),
with similar aims, coverage, and limitations. Indispensable for finding a particular compound, its properties, and references to it. Seven volumes, an
acute commentary on the growth of the field since 1950, when one 350-page
volume sufficed. **(Kosolapoff and Maier)**.

Contents are as follows (with approximate number of pages).

Volume 1 (1972). L. Maier, "Primary, Secondary, and Tertiary Phosphines,"
280 pages, literature to 1970. "Organophosphorus-metal compounds, biphos-
phines, triphosphines, tetraphosphines, cyclopolyphosphines, and corresponding oxides, sulfides, and selenides." 140 pages, literature to 1970.
G. Booth, "Phosphine Complexes with Metals," 110 pages, literature through
1969.

Volume 2 (1972). J. G. Verkade and K. J. Coskran, "Phosphite, Phosphonite, and Aminophosphine Complexes." 180 pages, literature to 1970.

P. Beck, "Quaternary Phosphonium Compounds," 320 pages, literature to 1970.

Volume 3 (1972). H. J. Bestmann and R. Zimmermann, "Phosphine Alkylenes and Other Phosphorus Ylids." 180 pages, literature to 1970.

D. Hellwinkel, "Penta-and Hexa-Organophosphorus Compounds." 150 pages, literature to 1970.

H. R. Hays and D. J. Peterson, "Tertiary Phosphine Oxides." 160 pages, literature to 1970.

Volume 4 (1972). L. Maier, "Tertiary Phosphine Sulfides, Selenides, and Tellurides." 74 pages, literature to May 1970.

M. Fild and R. Schmutzler, "Halo- and Pseudohalophosphines." 80 pages, literature through 1969.

M. Fild, R. Schmutzler, and S. C. Peake, "PHosphonyl- (Thiono-, Seleno-) and Phosphinyl- (Thiono-, Seleno-) Halides and Pseudohalides." 100 pages, literature through 1969.

A. Frank, "Phosphonous Acids (Thio-, Seleno-, analogs) and Derivatives." 200 pages, literature to 1970.

L. A. Hamilton and P. S. Landis, "Phosphinous Acids and Derivatives." 70 pages, literature to 1970.

Volume 5 (1973). M. Baudler, "Organic Derivatives of Hypophosphorous, Hypodiphosphorous, and Hypophosphoric Acids." 20 pages, literature to 1971.

W. Gerrard and H. R. Hudson, "Organic Derivatives of Phosphorous Acid and Thiophosphorous Acid." 300 pages, literature to 1971.

Volume 6 (1973). P. C. Crofts, "Phosphinic Acids and Derivatives." 210 pages, literature to 1971.

E. Cherbuliez, "Organic Derivatives of Phosphoric Acid." 360 pages, literature to 1971.

E. Fluck and W. Hambold, "Phosphorus(V)–Nitrogen Compounds with Phosphorus in Coordination Number 4." 250 pages, literature to 1971.

R. Keat and R. A. Shaw, "Cyclophosphazenes and Related Ring Compounds." 100 pages, literature to 1971.

Volume 7 (1976). K. H. Worms and M. Schmidt-Dunker, "Phosphonic Acids and Derivatives." 486 pages, literature to 1971 with some 1972 and 1973 references.

D. E. Ailman and R. J. Magee, "Organic Derivatives of Thio (Seleno, Telluro) Phosphoric Acid." 380 pages, literature to 1971 with some 1972 and 1973 references.

This volume also contains a complete listing of chapters for all the earlier volumes, plus errata.

F. G. Mann, *Heterocyclic Derivatives of Phosphorus, Arsenic, Antimony, and Bismuth*, Interscience, New York, 1970. Coverage includes only those ring systems with at least one carbon atom in the ring. The section on phosphorus of 350 pages covers the literature to 1969 and is very thorough. The book's emphasis is more on preparation and chemical properties than on physical properties or structure.

J. W. Mellor, *A Comprehensive Treatise of Inorganic and Theoretical Chemistry, Volume VIII, Supplement III, Phosphorus,* Longmans, London, 1971. More than 1400 pages giving a comprehensive view of the chemistry of phosphorus. Includes analytical aspects and a "chapter" of more than 200 pages on organic phosphorus chemistry. Also contains useful material on natural occurrence of phosphorus compounds and on analytical aspects. Fully referenced. Literature coverage to 1967. Excellent place to look for a detailed description (not too critical) of any area of inorganic phosphorus chemistry. (Mellor)

E. Muller, ed., *Methoden der Organischen Chemie (Houben-Weyl)*, Georg Thieme Verlag, Stuttgart. Volume 12/1, K. Sasse, *Organophosphorus Compounds* (1963). Discusses preparation, in particular (with many detailed instructions), and properties of compounds containing PC bonds; 670 pages with literature coverage to the end of 1961, and a very detailed index. A model of coverage and accuracy. Volume 12/2, K. Sasse, does the same for compounds without a PC bond, in 1100 pages, with literature coverage to the end of 1962. (Houben-Weyl).

Specialist Periodical Reports. Spectroscopic Properties of Inorganic and Organometallic Compounds. The Chemical Society, London. A continuing annual survey including all important spectroscopic analyses in a range of spectroscopic techniques. Valuable for finding pertinent literature. Volume 7 covers 1973, vol. 6, 1972, and so on, back to vol. 1 for 1967. (Spec. Props.)

I. O. Sutherland, ed., *Comprehensive Organic Chemistry*, vol. 2, Pergamon Press, Oxford, 1979, contains a minitreatise on organic compounds of phosphorus including C.N.3 (100 pages), C.N.4 (40 pages), and C.N.5 (25 pages), with an excellent discussion of ligand permutation, and phosphazenes and ylids (25 pages). Literature to 1977.

L. C. Thomas, *The Identification of Functional Groups in Organophosphorus Compounds,* Academic Press, London, 1974. A useful compact survey (120 pages) of analytical methods with particular stress on infrared spectroscopy and proton nmr. Contains many useful tables and chapter bibliographies. There is, inevitably, sustantial overlap with Thomas' recent book on infrared spectra of organophosphorus compounds (see Suggestions for Further Reading for Section 3.1).

A. D. F. Toy, *Phosphorus Chemistry in Everyday Living*, American Chemical Society, Washington, D.C., 1976. A nicely written general account of the ways in which phosphorus compounds are utilized in the home, in industry, in agriculture, and in life. The presentation of the chemical background is designedly at a simple level, so that the book can be understood even by a chemical novice.

S. Trippett, ed., *Specialist Periodical Reports. Organophosphorus Chemistry,* The Chemical Society, London. A continuing series of extreme utility. Although not easy to read because of the telescoped style (familiar to browsers in the Chemical Society's *Annual Reports*), these annual compendia provide an unrivaled guide to the subject literature. So far nine volumes have appeared, each covering the literature for one year (thus vol. 9 covers June 1976 to June 1977, vol. 8 covers June 1975 to June 1976, etc.). Each volume is about 280 pages long and appears within a year of the end of the period covered. Much inorganic chemistry is included where pertinent, despite the title. Has only an author index, but chapters are logically arranged and material is generally easy to locate.

J. R. Van Wazer, *Phosphorus and Its Compounds,* Interscience, New York, 1958. Volume 1 (chemistry) has some 900 pages on the systematics of phosphorus chemistry. This book represents an attempt (largely successful) to demonstrate that much of the chemistry of phosphorus can be treated in the unified way that the chemistry of carbon is in works on organic chemistry. An important work to read for its principles. Literature to 1957, so by now not at all current. Volume 2 (technology, biology, and applications), has 1100 pages and is more conventional and less successful than vol. 1. By now it is also somewhat out of date. (**Van Wazer**)

B. J. Walker, *Organophosphorus Chemistry*, Penguin Books, London, 1972. A brief survey, covering most significant topics; not overly critical. A handy introductory volume with good references for further reading covering the important literature up to 1970.

Index